A Short History of Mathematical Population Dynamics

T0215363

Nicolas Bacaër

A Short History
of Mathematical
Population
Dynamics

 Springer

Nicolas Bacaër
IRD (Institut de Recherche pour le Développement)
Bondy
France
nicolas.bacaer@ird.fr

Whilst we have made considerable efforts to contact all holders of copyright material contained in this book, we have failed to locate some of them. Should holders wish to contact the Publisher, we will make every effort to come to some arrangement with them

ISBN 978-0-85729-114-1   e-ISBN 978-0-85729-115-8
DOI 10.1007/978-0-85729-115-8
Springer London Dordrecht Heidelberg New York

British Library Cataloguing in Publication Data
A catalogue record for this book is available from the British Library

Mathematics Subject Classification (2010): 00A09, 01A05, 92D10, 92D25, 92D30, 92D40

Cover design: deblik

Printed on acid-free paper

Springer is part of Springer Science+Business Media (www.springer.com)

To Aili, Nina and Sophie

Preface

Population dynamics is the area of science which tries to explain in a simple mechanistic way the time variations of the size and composition of biological populations, such as those of humans, animals, plants or microorganisms. It is related to, but still quite distinct from, the more descriptive area of population statistics. One common point is that they make extensive use of mathematical language.

Population dynamics is at the intersection of various fields: mathematics, social sciences (demography), biology (population genetics and ecology) and medicine (epidemiology). As a result it is not often presented as a whole, despite the similarities between the problems met in various applications. A notable exception in French is the book *Mathematical Population Theories*[1] by Alain Hillion. But it presents the subject from the point of view of the mathematician, distinguishing various types of model: discrete-time models ($t = 0, 1, 2...$) and continuous-time models (t is a real number), deterministic models (future states are known exactly if the present state is known exactly) and stochastic models (where probabilities play a role). The book then considers logically discrete deterministic models, continuous deterministic models, discrete stochastic models and continuous stochastic models.

In the present book I have tried to discuss the same subject but from a historical point of view. Research is explained in its context. Short biographies of scientists are included. This should make the book easier to read for those less familiar with mathematics and can usually help in understanding the origin of the problems under study. But this book is not just about history. It can also serve as an introduction to mathematical modelling. It seemed important to include the details of most computations so that the reader can really see the limitations of the models. Technical parts are emphasized in grey boxes and can be skipped at first reading. The last chapter focuses on the numerous contemporary problems in population dynamics that one can try to analyze from a mathematical point of view. For those who would like to know more, the lists of references at the end of each chapter also include web sites from which original articles may be downloaded.

---

[1] Presses Universitaires de France, Paris, 1986. For English readers, *Quantitative Methods in Biological and Medical Sciences* by H.O. Lancaster (Springer-Verlag, 1994) gives a good historical overview without the mathematical details.

It was not possible in a book of this length to give a complete picture of all the work developed until now or to talk about all the scientists who have contributed to the subject. The choice made necessarily contains an arbitrary component, particularly for the most recent decades. I hope nevertheless that the sample chosen is representative enough, and that people active in the field whose work is not mentioned will not be upset.

The ideal audience for this book would include:

- High school and university students wondering what links may exist between the mathematics courses they have to attend and the world around them, or students preparing personal work on a theme related to population dynamics.
- Mathematics teachers trying to make their course more attractive. The knowledge of the four elementary operations is enough to understand most of Chapters 1, 2 and 5. Chapter 3 can serve as an introduction to the applications of logarithms. This book also covers: recurrence equations in Chapters 1, 3, 8, 11, 14, 21, 23, 24; differential equations in Chapters 4, 6, 12, 13, 16; partial differential equations in Chapters 20, 25; an integral equation in Chapter 10; and applications of probability theory in Chapters 2, 7, 8, 9, 15, 16, 17, 18, 19, 22.
- People already familiar with demography, epidemiology, genetics or ecology and willing to compare their favourite area with others which may involve similar mathematical models.
- Readers interested in the history of science.

This book is essentially a translation of the French edition published by Cassini Éditeurs (Paris) in 2008 under the title *Histoires de Mathématiques et de Populations*. Some chapters have been reorganized or rewritten. Four figures have been added. A few misprints have been corrected. The lists of references at the end of each chapter have been extended and updated. These lists include web sites showing the original works.

A reference followed by a URL means that it can be easily found – at least in part – by searching on the World Wide Web (e.g. through Google Books).

A number of people have made remarks on various versions of the book, provided references and pictures or discussed copyright issues: André and Catherine Bellaïche, Bernard Bru, Joe Gani, Geoffrey Grimmett, François Hamel, Nikolai Nikolski, Carel Pretorius and Niklaus Salzmann. Thanks to all of them. I also thank my mother for helping with the translation.

Marrakech, August 2010
Nicolas Bacaër

Contents

Chapter 1
The Fibonacci sequence (1202)

Leonardo of Pisa, named Fibonacci long after his death, was born *circa* 1170 in the Republic of Pisa when it was at the height of its commercial and military power in the Mediterranean world. Around 1192, Fibonacci's father was sent by the Republic to the harbor of Bejaia, now in Algeria, to head a trading post. His son joined him shortly after to prepare to be a merchant. Leonardo started to learn the decimal number system that the Arabs had brought back from India and which is still in use today in almost the same form: 0, 1, 2, 3, 4, 5, 6, 7, 8 and 9. While traveling for business around the Mediterranean Sea, he compared the different number systems and studied Arab mathematics. Back in Pisa, he finished writing in 1202 a book in Latin entitled *Liber abaci* (Book of Calculation) in which he explained the new number system and showed how to use it for accounting, weight and currency conversions, interest rates and many other applications. He also gathered most of the results in algebra and arithmetic known to the Arabs.

Fibonacci considered in his book what one would today call a problem in population dynamics. But it appeared just as a computational exercise in the middle of other unrelated subjects: the preceding section in the book is about perfect numbers that are the sum of their factors, like $28 = 14 + 7 + 4 + 2 + 1$, and the following section is a problem about the sharing of money among four people that is equivalent to a linear system of four equations. Here is a translation from Latin of the population problem:

> A certain man had one pair of rabbits together in a certain enclosed place. One wishes to know how many are created from the pair in one year when it is the nature of them in a single month to bear another pair and in the second month those born to bear also.

If there is a pair of newborn rabbits at the beginning of the first month, this pair will not yet be fertile after one month and there will still be just one pair of rabbits at the beginning of the second month. This pair of rabbits will give birth to another pair at the beginning of the third month, so there will be two pairs in total. The initial pair of rabbits will again give birth to another pair at the beginning of the fourth month. But the second pair of rabbits will not yet be fertile. There will be just three pairs of rabbits.

N. Bacaër, *A Short History of Mathematical Population Dynamics*,
DOI 10.1007/978-0-85729-115-8_1, © Springer-Verlag London Limited 2011

Using modern notations let P_n be the number of pairs of rabbits at the beginning of month n. The number of pairs of rabbits P_{n+1} in month $n+1$ is the sum of the number P_n of pairs in month n and of the number of newborn pairs in month $n+1$. But only the pairs of rabbits that are at least two months old give birth to new pairs of rabbits in month $n+1$. These are the pairs that were already there in month $n-1$ and their number is P_{n-1}. So

$$P_{n+1} = P_n + P_{n-1}.$$

This is a recurrence relation: it gives the population in month $n+1$ as a function of the population in the previous months. Hence Fibonacci could easily build the following table, where $1+1=2$, $1+2=3$, $2+3=5$, $3+5=8$, etc.

n	1	2	3	4	5	6	7	8	9	10	11	12	13
P_n	1	1	2	3	5	8	13	21	34	55	89	144	233

In fact, Fibonacci considered as the initial condition the situation in month $n=2$. Since $P_{14} = 144 + 233 = 377$, he finally obtained 377 pairs of rabbits twelve months after his starting point. He noticed that this sequence of numbers could continue indefinitely.

After 1202 Fibonacci wrote several other books, such as *Practica geometriae* in 1220 and *Liber quadratorum* (Book of Squares) in 1225. His reputation led to a meeting with the emperor Frederick II, who appreciated science. In 1240 the Republic of Pisa awarded Fibonacci a yearly pension. The year of his death is unknown.

During the following centuries, Fibonacci's rabbit problem was forgotten and had no influence on the development of mathematical models for population dynamics. Several scientists met the same sequence of numbers in their studies but did not refer to Fibonacci or to any population. Several of Kepler's books contain the remark that the ratio P_{n+1}/P_n converges, when n tends to infinity, to the golden number $\phi = (1+\sqrt{5})/2$. This is a particular case of a property common to most population models: the tendency to increase geometrically (see Chapters 3 and 21). In 1728 Daniel Bernoulli obtained the exact formula

$$P_n = \frac{1}{\sqrt{5}} \left[\frac{1+\sqrt{5}}{2} \right]^n - \frac{1}{\sqrt{5}} \left[\frac{1-\sqrt{5}}{2} \right]^n$$

while studying general recurrent series. The complete works of Fibonacci were published in the nineteenth century. From then on, the sequence (P_n) could be found in books of recreational mathematics under the name of Fibonacci sequence.

It is clear that, in order to model a population of rabbits, the hypotheses leading to the Fibonacci sequence are far from being realistic: no mortality, no separation of sexes etc. Our interest in this sequence in recent decades in biology has come from the fact that several plants contain structures that involve some of the numbers P_n, for example, 8 and 13 in pine cones or 34 and 55 in sunflowers. A scientific journal, *The Fibonacci Quarterly*, is even entirely dedicated to the properties and applications of the Fibonacci sequence!

Further reading

1. Bernoulli, D.: Observationes de seriebus... *Comment. Acad. Sci. Imp. Petropolitanae* **3**, 85–100 (1728/1732) Reproduced in: *Die Werke von Daniel Bernoulli*, Band 2, Birkhäuser, Basel, 1982, pp. 49–64 books.google.com
2. Sigler, L.E.: *Fibonacci's Liber Abaci*. Springer, New York (2002) books.google.com
3. Vogel, K.: Leonardo Fibonacci. In: Gillespie, C.C. (ed.) *Dictionary of Scientific Biography*, vol. 4, pp. 604–613. Scribner, New York (1971)

Chapter 2
Halley's life table (1693)

Edmond Halley was born near London in 1656. His father was a rich soap maker. Edmond became interested in astronomy at a young age. He started studying at Queen's College of Oxford University. When the Greenwich Observatory was inaugurated in 1675, Halley could already visit Flamsteed, the Astronomer Royal. He interrupted his studies from 1676 to 1678 to go to the island of Saint Helena and establish a catalog of the stars that can be seen from the southern hemisphere. At his return to England he became a fellow of the Royal Society. He published also the observations he had made on the circulation of winds during his journey to Saint Helena. In 1684 he visited Newton in Cambridge to discuss the link between Kepler's laws of planetary motion and the force of attraction exerted by the Sun. He encouraged Newton to write the famous *Mathematical Principles of Natural Philosophy*, a book which he finally published at his own expense. He was then working as clerk of the Royal Society. In 1689 he designed a bell for underwater diving, which he tested himself.

Fig. 2.1 Edmond Halley
(1656–1742)

N. Bacaër, *A Short History of Mathematical Population Dynamics*,
DOI 10.1007/978-0-85729-115-8_2, © Springer-Verlag London Limited 2011

At about the same time, Caspar Neumann, a theologian living in Breslau, was collecting data about the number of births and deaths in his city. Breslau belonged to the Habsburg empire (it is now in Poland and called Wrocław). The data included the age at which people had died. So it could be used to construct a life table showing the probability of surviving until any given age.

The first life table had been published in London in 1662 in a book entitled *Natural and Political Observations Made upon the Bills of Mortality*. This book is usually considered as the founding text of both statistics and demography and has a strange particularity: people still wonder nowadays if it was written by John Graunt, a London merchant and author indicated on the book cover, or by his friend William Petty, one of the founders of the Royal Society[1]. In any case the life table contained in the book tried to take advantage of the bulletins that had been regularly reporting the burials and baptisms in London since the beginning of the seventeenth century. These bulletins were mainly used to inform people on the recurrent epidemics of plague. This is the reason why they indicated the cause of death and not the age at which people died. To obtain a life table giving the chance of survival as a function of age, Graunt or Petty had to guess how different causes of death were related to age groups. So their life table could be subject to large errors. The book was nevertheless very successful, with five editions between 1662 and 1676. Several cities in Europe had started to publish bulletins similar to that of London.

So it was nearly thirty years after this first life table that, following the suggestion of Leibniz, Neumann sent to Henry Justel, the secretary of the Royal Society, his demographic data from the city of Breslau for the years 1687–1691. Justel died shortly after, and Halley got hold of the data, analyzed them and in 1693 published his conclusions in the *Philosophical Transactions of the Royal Society*. His article is called "An estimate of the degrees of the mortality of mankind, drawn from curious tables of the births and funerals at the city of Breslaw, with an attempt to ascertain the price of annuities upon lives".

For the period of five years under study, Halley noticed that the number of births in Breslau was more or less equal to the number of deaths, so that the total population was almost constant. To simplify the analysis, he assumed that the population was exactly at steady state: the annual number of births (call it P_0), the total population, the population aged k (P_k) and the annual number of deaths at age k (D_k) are all constant as time goes by. This emphasizes an additional interesting property of the data from Breslau, because such a simplification would not have been possible for a fast growing city such as London, where the statistics were also biased by the flow of population coming from the countryside.

The data from Breslau had a mean of $1,238$ births per year: this is the value that Halley took for P_0. In principle he could also compute from the data the annual mean D_k of the number of deaths among people aged k for all $k \geq 0$. Using the formula

$$P_{k+1} = P_k - D_k, \qquad (2.1)$$

---

[1] For a detailed discussion, see the book by Hervé Le Bras in the references.

Table 2.1 Halley's life table showing the population P_k aged k.

Age	Number	Age	Number	Age	Number
1	1000	29	539	57	272
2	855	30	531	58	262
3	798	31	523	59	252
4	760	32	515	60	242
5	732	33	507	61	232
6	710	34	499	62	222
7	692	35	490	63	212
8	680	36	481	64	202
9	670	37	472	65	192
10	661	38	463	66	182
11	653	39	454	67	172
12	646	40	445	68	162
13	640	41	436	69	152
14	634	42	427	70	142
15	628	43	417	71	131
16	622	44	407	72	120
17	616	45	397	73	109
18	610	46	387	74	98
19	604	47	377	75	88
20	598	48	367	76	78
21	592	49	357	77	68
22	586	50	346	78	58
23	579	51	335	79	50
24	573	52	324	80	41
25	567	53	313	81	34
26	560	54	302	82	28
27	553	55	292	83	23
28	546	56	282	84	20

he could construct Table 2.1 giving P_k. Conversely, one can find the values of D_k that he used from the formula $D_k = P_k - P_{k+1}$: $D_0 = 238$, $D_1 = 145$, $D_2 = 57$, $D_3 = 38$ and so on. In fact, Halley rearranged his results a little, either to get round numbers (this is the case of D_1, which has been slightly changed so that $P_1 = 1,000$) or to smooth certain irregularities due to the small numbers of deaths at old ages in a five-year study. Taking the sum of all the numbers P_k in the table, Halley obtained an estimate of the total population of Breslau close[2] to 34,000. In summary this method had the great advantage of not requiring a general census but only knowledge of the number of births and deaths and of the age at which people died during a few years.

Halley's life table served as a reference for various works in the eighteenth century (see Chapter 4). Indeed, although the values of P_k were specific to the city of Breslau, one could consider that the ratio P_{k+1}/P_k was the probability of surviving until age $k + 1$ knowing that one had already reached age k. This probability could reasonably be used for the populations of other European cities

---

[2] For people aged over 84 years, Halley just mentioned that their number was 107.

of the time. For example, one might expect a one-year-old child to have 661 chances out of 1,000 of reaching age 10 or 598 chances out of 1,000 of reaching age 20.

Halley also used his life table to compute the price of annuities upon lives. During the sixteenth and seventeenth centuries, several cities and states had sold such annuities to their citizens to raise money. The buyers received each year until their death a fixed amount of money, which was equal to a certain percentage of the sum initially paid, often twice the interest rate of the time, but independently of the age of the buyer. Of course the institution was risking bankruptcy if too many people with a very long life expectancy bought these annuities. The problem could not be correctly addressed without a reliable life table.

In 1671 Johan De Witt, prime minister of Holland, and Johannes Hudde, one of the mayors of the city of Amsterdam, had already thought about the problem of computing the price of life annuities. Fearing an invasion of French troops, they wanted to raise money to strengthen the army. They had data concerning people who had bought annuities upon lives several decades earlier, in particular the age at which the annuities had been bought and the age at which people had died. They had managed to compute the price of annuities more or less correctly, but their method was later forgotten. Holland was invaded the following year and De Witt was lynched by the crowd.

Halley considered the problem anew in 1693 with the life table from Breslau and assuming an interest rate of 6%. The method of computation is simple. Let i be the interest rate. Let R_k be the price at which a person aged k can buy an annuity of, say, one pound per year. This person has a probability P_{k+n}/P_k of being still alive at age $k+n$. The pound that the State promises to pay if he reaches this age can be obtained by placing $1/(1+i)^n$ pounds of the initial sum at the interest rate i. So if one makes the simplifying assumption that the initial sum is used only to pay the annuities, then the price should be

$$R_k = \frac{1}{P_k}\left(\frac{P_{k+1}}{1+i} + \frac{P_{k+2}}{(1+i)^2} + \frac{P_{k+3}}{(1+i)^3} + \cdots\right). \qquad (2.2)$$

Halley obtained in this way Table 2.2, which shows the factor R_k by which the desired annuity has to be multiplied to get the necessary initial sum. A man aged 20 would hence get each year $1/12.78 \simeq 7.8\%$ of the initial sum. But a man aged 50 would get $1/9.21 \simeq 10.9\%$, because he would have fewer years to live. Notice that twice the interest rate would correspond to an annuity equal to 12% of the initial sum, or equivalently to a price equal to 8.33 times the annuity.

The computations are of course quite tedious. Halley could nevertheless use tables of logarithms to obtain the general term $P_{k+n}/(1+i)^n$ more quickly. Since he did not show values for P_k above 84 years, it is not possible to check his calculations exactly. Finally, Halley's work did not have any immediate impact: for several decades, annuities upon lives in England and elsewhere continued to be sold at a price independent of the age of the buyer and at a price that was much lower than it could be, for example 7 times the annuity.

Table 2.2 Multiplying factor giving the price of annuities upon lives.

Age k	Price R_k	Age k	Price R_k	Age k	Price R_k
1	10.28	25	12.27	50	9.21
5	13.40	30	11.72	55	8.51
10	13.44	35	11.12	60	7.60
15	13.33	40	10.57	65	6.54
20	12.78	45	9.91	70	5.32

The questions derived from life tables interested many scientists during Halley's time. The Dutch Christiaan Huygens, author in 1657 of the first booklet dedicated to probability theory, discussed in 1669 in his correspondence with his brother Graunt's life table and the calculation of life expectancy[3]. A few years before bringing Neumann into contact with the Royal Society, Leibniz also wrote about the calculation of life expectancy in an essay which remained unpublished. In 1709 it was the turn of Nikolaus I Bernoulli. In 1725 Abraham de Moivre published an entire *Treatise on Annuities*. He noticed in particular that the price R_k could be easily computed for old ages since formula (2.2) contained just a few terms. One could then use the backward recurrence formula

$$R_k = \frac{P_{k+1}}{P_k} \frac{1 + R_{k+1}}{1 + i},$$

which is easily proved starting from (2.2). Using the value that Halley gives for the price at age 70, one can hence check[4] the other values of Table 2.2.

After this break focusing on demography Halley returned to his main research subjects. Between 1698 and 1700 he sailed around the Atlantic Ocean to draw a map of the Earth's magnetic field. In 1704 he became professor at Oxford University. The following year he published a book on comets and predicted that the comet of 1682, which Kepler had observed in 1607, would come back in 1758: it became known as "Halley's comet". He also published a translation of the book by Apollonius of Perga on conics. In 1720 he replaced Flamsteed as Astronomer Royal. He tried to solve the problem of determining longitude at sea precisely from observation of the Moon, a problem of great practical importance for navigation. He died in Greenwich in 1742 at age 86.

Further reading

1. Fox, M.V.: *Scheduling the Heavens: The Story of Edmond Halley*. Morgan Reynolds, Greensboro, North Carolina (2007)

---

[3] The life expectancy at age k is given by formula (2.2) with $i = 0$.

[4] It seems that there are a few errors in the table, in particular for the ages 5 and 15.

2. Graunt, J.: *Natural and Political Observations Mentioned in a Following Index and Made upon the Bills of Mortality*, 3rd edn. London (1665). `echo.mpiwg-berlin.mpg.de`
3. Hald, A.: *A History of Probability and Statistics and Their Applications before 1750*. Wiley, Hoboken, New Jersey (2003). `books.google.com`
4. Halley, E.: An estimate of the degrees of the mortality of mankind, drawn from curious tables of the births and funerals at the city of Breslaw; with an attempt to ascertain the price of annuities upon lives. *Phil. Trans. Roy. Soc. London* **17**, 596–610 (1693). `gallica.bnf.fr`
5. Heyde, C.C.: John Graunt. In: Heyde, C.C., Seneta, E. (eds.) *Statisticians of the Centuries*, pp. 14–16. Springer, New York (2001)
6. Koch, P.: Caspar Neumann. In: Heyde, C.C., Seneta, E. (eds.) *Statisticians of the Centuries*, pp. 29–32. Springer, New York (2001)
7. Le Bras, H.: *Naissance de la mortalité, L'origine politique de la statistique et de la démographie*. Gallimard, Paris (2000)

Chapter 3
Euler and the geometric growth of populations (1748–1761)

Leonhard Euler was born in 1707 in Basel, Switzerland. His father was a Protestant minister. In 1720 Euler started studying at the university. He also received private mathematics lessons from Johann Bernoulli, one of the most famous mathematicians of the generation after Leibniz and Newton. He made friends with two of Johann Bernoulli's sons: Nikolaus II and Daniel. In 1727 Euler joined Daniel at the newly created Academy of Sciences in Saint Petersburg. Apart from mathematics he was also interested in physics and many other scientific and technical subjects. In 1741 King Frederick II of Prussia invited him to become the director of the mathematics section of the Academy of Sciences in Berlin. Euler published a considerable number of articles and books on all aspects of mechanics (astronomy, elasticity, fluids, solids) and mathematics (number theory, algebra, infinite series, elementary functions, complex numbers, differential and integral calculus, differential and partial differential equations, optimization, geometry) but also on demography. He was the most prolific mathematician of his time.

Fig. 3.1 Euler (1707–1783)

N. Bacaër, *A Short History of Mathematical Population Dynamics*,
DOI 10.1007/978-0-85729-115-8_3, © Springer-Verlag London Limited 2011

In 1748 Euler published a treatise in Latin entitled *Introduction to Analysis of the Infinite*. He considered six examples in the chapter on exponentials and logarithms: one on the mathematical theory of musical scales, another on the repayment of a loan with interest, and four on population dynamics. In the latter Euler assumed that the population P_n in year n satisfies

$$P_{n+1} = (1+x)P_n$$

for all integer n. The growth rate x is a positive real number. Starting from an initial condition P_0, the population in year n is given by

$$P_n = (1+x)^n P_0 .$$

This is called geometric or exponential growth. The first example asks:

> If the population in a certain region increases annually by one thirtieth and at one time there were 100,000 inhabitants, we would like to know the population after 100 years.

The answer is $P_{100} = (1+1/30)^{100} \times 100,000 \simeq 2,654,874$. For this example Euler was inspired by the census of Berlin that took place in 1747 and which yielded an estimate of 107,224 for the population. His calculation shows that a population can increase more than tenfold within one century. This is precisely what had been observed at the time for the city of London.

One should note that computing $(1+1/30)^{100}$ is very easy with a modern pocket calculator. But in Euler's time one had to use logarithms to avoid numerous multiplications by hand and get the result rapidly. One computes first the decimal logarithm (in base 10) of P_{100}. The fundamental property of the logarithm $\log(ab) = \log a + \log b$ shows that

$$\log P_{100} = 100 \log(31/30) + \log(100,000) = 100 (\log 31 - \log 30) + 5 .$$

Logarithms had been introduced in 1614 by the Scotsman John Napier. His friend Henry Briggs had published the first table of decimal logarithms in 1617. In 1628 the Dutch Adriaan Vlacq had completed Briggs' work by publishing a table containing decimal logarithms of integers from 1 to 100,000 with ten-digit precision. This is the kind of table that Euler used to get $\log 30 \simeq 1.477121255$, $\log 31 \simeq 1.491361694$, and finally $\log P_{100} \simeq 6.4240439$. It remains to find the number P_{100} whose logarithm is known. Since decimal logarithms of integers from 1 to 100,000 range from 0 to 5, one looks instead for the logarithm of $P_{100}/100$, which is 4.4240439. One can check in the table of logarithms that $\log 26,548 \simeq 4.424031809$ and $\log 26,549 \simeq 4.424048168$. Replacing the logarithmic function by a straight line between 26,548 and 26,549, Euler obtained that

$$\frac{P_{100}}{100} \simeq 26,548 + \frac{4.4240439 - 4.424031809}{4.424048168 - 4.424031809} \simeq 26,548.74 .$$

So $P_{100} \simeq 2,654,874$.

The second example concerning population dynamics in Euler's book is as follows:

> Since after the Flood all men descended from a population of six, if we suppose that the population after two hundred years was 1,000,000, we would like to find the annual rate of growth.

Since $10^6 = (1+x)^{200} \times 6$, we get with a pocket calculator $x = (10^6/6)^{1/200} - 1 \simeq 0.061963$. With tables of logarithms one has to go through $\log(10^6) = 200 \log(1 + x) + \log 6$ to get $\log(1+x) = (6 - \log 6)/200 \simeq 0.0261092$ and $1 + x \simeq 1.061963$. Thus Euler could conclude that the population would increase by $x \simeq 1/16$ per year. To understand the origin of this example, one has to remember that contemporary philosophers were starting to deny the truth of biblical stories. A literal reading would fix the time of the Flood around 2350 BC with the following survivors: Noah, his three sons and their wives. The book of Genesis says:

> These three were the sons of Noah; and from these the whole earth was peopled.

A population growth rate of $1/16$ (or 6.25%) per year after the Flood did not seem too unrealistic to Euler. Being the son of a Protestant minister and having stayed religious all his life, he concluded:

> For this reason it is quite ridiculous for the incredulous to object that in such a short space of time the whole earth could not be populated beginning with a single man[1].

Euler also noticed that if the growth had continued at the same pace until 400 years after the Flood, the population would have been $(1+x)^{400} \times 6 = (10^6/6)^2 \times 6 \simeq 166$ billion:

> However, the whole earth would never be able to sustain that population.

This idea would be greatly developed by Malthus half a century later (see Chapter 5).

Euler's third example asks:

> If each century the human population doubles, what is the annual rate of growth?

Since $(1+x)^{100} = 2$, we get with a pocket calculator $x = 2^{1/100} - 1 \simeq 0.00695$. With tables of logarithms, $100 \log(1+x) = \log 2$. So $\log(1+x) \simeq 0.0030103$ and $1 + x \simeq 1.00695$. Hence the population grows by $x \simeq 1/144$ each year. The fourth and last example asks in the same way:

> If the human population increases annually by 1/100, we would like to know how long it will take for the population to become ten times as large.

---

[1] In the book published by Graunt in 1662 (see Chapter 2), one finds a similar remark:

> One couple viz. Adam and Eve, doubling themselves every 64 years of the 5,160 years, which is the age of the world according to the Scriptures, shall produce far more people, than are now in it. Wherefore the world is not above 100 thousand years old, as some vainly imagine, nor above what the Scripture makes it.

With $(1+1/100)^n = 10$, we find $n \log(101/100) = 1$. So $n = 1/(\log 101 - 2) \simeq 231$ years. This is all that can be found in the *Introduction to Analysis of the Infinite* from 1748 concerning population dynamics. Euler would come back to this subject more thoroughly some years later.

In 1760 he published in the proceedings of the Academy of Sciences in Berlin a work entitled "A general investigation into the mortality and multiplication of the human species". This work was a kind of synthesis between his previous analysis of the geometric growth of populations and earlier studies on life tables (see Chapter 2). Euler considered for example the problem:

> Knowing the number of births and burials which happen during the course of one year, to find the number of all the living and their annual increase, for a given hypothesis of mortality.

Euler assumed here that the following numbers are known:

- the number of births B_n during year n;
- the number of deaths D_n during year n;
- the proportion q_k of newborns that reach[2] age $k \geq 1$.

Let P_n be the population[3] in year n. Euler made two extra implicit assumptions:

- the population increases geometrically: $P_{n+1} = r P_n$ (we set $r = 1 + x$);
- the ratio between births and population is constant: $B_n/P_n = m$.

These two assumptions imply that the number of births increases geometrically and at the same rate: $B_{n+1} = r B_n$. Euler then considered the state of the population at hundred-year interval, say between the years $n = 0$ and $n = 100$, assuming that nobody survives beyond hundred years. To clarify the presentation, call $P_{k,n}$ $(k \geq 1)$ the population alive at the beginning of year n, which was born in the year $n - k$. Call $P_{0,n} = B_n$ the number of births during year n. From the definition of the survival coefficient q_k, we have $P_{k,n} = q_k P_{0,n-k} = q_k B_{n-k}$. So

$$
\begin{aligned}
r^{100} P_0 = P_{100} &= P_{0,100} + P_{1,100} + \cdots + P_{100,100} \\
&= B_{100} + q_1 B_{99} + \cdots + q_{100} B_0 \\
&= (r^{100} + r^{99} q_1 + \cdots + q_{100}) B_0.
\end{aligned}
$$

Dividing this equation by $r^{100} P_0$, we obtain

$$
1 = m \left(1 + \frac{q_1}{r} + \frac{q_2}{r^2} + \cdots + \frac{q_{100}}{r^{100}} \right). \tag{3.1}
$$

This is the equation that is sometimes called "Euler's equation" in demography. Counting births and deaths separately, we get

---

[2] More precisely, that are still alive at the beginning of the year of their k^{th} birthday.

[3] In fact P_n is the number of people alive during at least part of the year n. This includes the people alive at the beginning of the year and people born during the year.

$$rP_n = P_{n+1} = P_n - D_n + B_{n+1} = P_n - D_n + rB_n. \qquad (3.2)$$

So the number of deaths also increases geometrically: $D_{n+1} = rD_n$. Moreover,

$$\frac{1}{m} = \frac{P_n}{B_n} = \frac{D_n/B_n - r}{1 - r}. \qquad (3.3)$$

Replacing this in equation (3.1), we arrive finally at the equation

$$\frac{D_n/B_n - 1}{1 - r} = \frac{q_1}{r} + \frac{q_2}{r^2} + \cdots + \frac{q_{100}}{r^{100}}, \qquad (3.4)$$

where there is only one unknown left: r. This is what is usually called an implicit equation because we cannot extract r as a function of the other parameters. But we can compute the left- and right-hand sides of equation (3.4) for a fixed value of r and let r vary until the two sides are equal. The value of r thus obtained gives the growth rate $x = r - 1$ of the population. Notice that from equations (3.1) and (3.3), we obtain for the population P_n the following expression:

$$P_n = B_n \left(1 + \frac{q_1}{r} + \frac{q_2}{r^2} + \cdots + \frac{q_{100}}{r^{100}} \right).$$

When the population is stationary ($r = 1$), this expression is the same as the one used by Halley to estimate the population of the city of Breslau (see Chapter 2).

Euler also considered the following question:

> The hypotheses of mortality and fecundity being given, if one knows the number of all the living, to find how many there will be at each age.

Since the survival coefficients q_k and the fertility coefficient m are known, the growth rate r can be computed from equation (3.1). During year n, the number of people born in year $n - k$ is $q_k B_{n-k} = q_k B_n / r^k$ (with $q_0 = 1$). So the proportion of the total population that is aged k is

$$\frac{q_k / r^k}{1 + q_1/r + q_2/r^2 + \cdots + q_{100}/r^{100}}.$$

This proportion is constant. Using Lotka's terminology (see Chapter 10), the population is said to be "stable": the age pyramid keeps the same shape through time.

Euler then reexamined the problem of constructing a life table when the population is not stationary but increases geometrically:

> Knowing the number of all the living, similarly the number of births with the number of deaths at each age during the course of one year, to find the law of mortality.

By "law of mortality", Euler meant the set of survival coefficients q_k. The total population is now assumed to be known through a census, which was not the case for Halley (see Chapter 2). Equation (3.2) shows that the growth rate is

$$r = \frac{P_n - D_n}{P_n - B_n}.$$

Let $D_{k,n}$ be the number of people who die at age k during year n: these people were born in the year $n-k$. So $D_{k,n} = (q_k - q_{k+1})B_{n-k}$. But $B_{n-k} = B_n/r^k$. The survival coefficients q_k can therefore be computed with the recurrence formula

$$q_{k+1} = q_k - \frac{r^k D_{k,n}}{B_n}$$

for all $k \geq 0$, with $q_0 = 1$. This formula multiplied by B_n gives back the formula (2.1) used by Halley for the stationary case $r = 1$. Euler insisted nevertheless on the fact that his method of computing the survival coefficients q_k assumes that the population increases regularly, excluding accidents such as plague epidemics, wars, famines etc. If the censuses in Euler's time had recorded the age of the people (as in Sweden), this assumption would have been unnecessary and the coefficients q_k could have been computed more easily.

Given the survival coefficients q_k, Euler also showed how to compute the price of annuities upon lives. He didn't mention the works of Halley or de Moivre on this subject. Euler used an interest rate of 5% and the life table published in 1742 by the Dutchman Willem Kersseboom.

Euler was not the only scientist interested in demography at the Berlin Academy. His colleague Johann Peter Süssmilch had published in 1741 a treatise in German entitled *The Divine Order in the Changes of the Human Generation, Through the Birth, the Deaths and the Procreation of the Same Established*, which is considered nowadays as the first treatise entirely dedicated to demography. Süssmilch had also written a book *On the Rapid Growth of the City of Berlin* in 1752.

Fig. 3.2 Süssmilch
(1707–1767)

In 1761 Süssmilch published a second edition of his treatise. In the chapter entitled "On the rate of increase and on the doubling time of populations", he included an interesting mathematical model that Euler had worked out for him. The model was similar to that of Fibonacci (see Chapter 1) but for a human population. Starting with a couple (one man and one woman) both aged 20 years in the year 0, Euler assumed that people die at the age of 40 and marry at the age of 20, while each couple has six children: two children (a boy and a girl) at the age of 22, another two at the age of 24 and the last two at the age of 26. Counting the years two by two so that B_i is the number of births during the year $2i$, Euler concluded that

$$B_i = B_{i-11} + B_{i-12} + B_{i-13} \tag{3.5}$$

for all $i \geq 1$. The initial conditions correspond to $B_{-12} = 0$, $B_{-11} = 0$, $B_{-10} = 2$ and $B_i = 0$ for $-9 \leq i \leq 0$. Euler could thus compute the number of births, as shown in the second column of Table 3.1. The number of deaths D_i in year $2i$ is then equal to the number of births in year $2i - 40$: $D_i = B_{i-20}$ for $i \geq 10$ while $D_i = 0$ for $i \leq 9$. As for the number P_i of people alive in year $2i$, it is equal to the number of people alive in year $2i - 2$, plus the number of births in year $2i$, minus the number of deaths in year $2i$: $P_i = P_{i-1} + B_i - D_i$. This chapter in Süssmilch's book ends with a remark that could have already been made about the Fibonacci sequence:

> The great disorder that seems to prevail in Euler's table does not prevent the number of births from following a kind of progression that one calls recurrent series [...] Whatever the initial disorder of these progressions, they turn into a geometric progression if they are not interrupted and the disorders of the beginning fade little by little and vanish almost completely.

The book does not say more about the mathematics of this population model. However, Euler pushed the study much further in a manuscript entitled "On the multiplication of the human race", which stayed unpublished during his lifetime. Looking for a solution of equation (3.5) of the form $B_i = c r^i$, i.e. of the form of a geometric progression, he obtained after simplification a polynomial equation of degree 13:

$$r^{13} = r^2 + r + 1. \tag{3.6}$$

He looked for a solution close to $r = 1$ and noticed, using a table of logarithms for the computation of r^{13}, that

$$1 + r + r^2 - r^{13} \simeq \begin{cases} 0.212 & \text{if } r = 1.09, \\ -0.142 & \text{if } r = 1.10. \end{cases}$$

So equation (3.6) has a root between 1.09 and 1.10. Approximating the function $1 + r + r^2 - r^{13}$ by a line segment on this interval, Euler obtained

$$r \simeq \frac{0.142 \times 1.09 + 0.212 \times 1.10}{0.142 + 0.212} \simeq 1.0960.$$

The years being counted two by two, the number of births tends to be multiplied by $\sqrt{r}$ each year. This number doubles every n years if $(\sqrt{r})^n = 2$, i.e. every $n = 2 \log 2 / \log r \simeq 15$ years. Since asymptotically $B_i \simeq c r^i$ and since the number D_i of deaths in year $2i$ is equal to B_{i-20}, we obtain $D_i \simeq B_i / r^{20}$ with $r^{20} \simeq 6.25$. The number of births is about six times the number of deaths. The number P_i of people

Table 3.1 Euler's table.

i	Births	Deaths	Living	i	Births	Deaths	Living	i	Births	Deaths	Living
0	0	0	2	40	20	0	206	80	86	180	5,584
1	2	0	4	41	8	0	214	81	226	252	5,558
2	2	0	6	42	2	0	216	82	532	282	5,808
3	2	0	8	43	0	2	214	83	1,008	252	6,564
4	0	0	8	44	0	6	208	84	1,568	180	7,952
5	0	0	8	45	2	12	198	85	2,032	100	9,884
6	0	0	8	46	10	14	194	86	2,214	42	12,056
7	0	0	8	47	30	12	212	87	2,032	14	14,074
8	0	0	8	48	60	6	266	88	1,568	16	15,626
9	0	0	8	49	90	2	354	89	1,010	56	16,580
10	0	2	6	50	102	0	456	90	550	154	16,976
11	0	0	6	51	90	0	546	91	314	322	16,968
12	2	0	8	52	60	0	606	92	384	532	16,820
13	4	0	12	53	30	0	636	93	844	714	16,950
14	6	0	18	54	10	2	644	94	1,766	786	17,930
15	4	0	22	55	2	8	638	95	3,108	714	20,324
16	2	0	24	56	2	20	620	96	4,608	532	24,400
17	0	0	24	57	12	32	600	97	5,814	322	29,892
18	0	0	24	58	42	38	604	98	6,278	156	36,014
19	0	0	24	59	100	32	672	99	5,814	72	41,756
20	0	0	24	60	180	20	832	100	4,610	86	46,280
21	0	2	22	61	252	8	1,076	101	3,128	226	49,182
22	0	2	20	62	282	2	1,356	102	1,874	532	50,524
23	2	2	20	63	252	0	1,608	103	1,248	1,008	50,764
24	6	0	26	64	180	0	1,788	104	1,542	1,568	50,738
25	12	0	38	65	100	2	1,886	105	2,994	2,032	51,700
26	14	0	52	66	42	10	1,918	106	5,718	2,214	55,204
27	12	0	64	67	14	30	1,902	107	9,482	2,032	62,654
28	6	0	70	68	16	60	1,858	108	13,530	1,568	74,616
29	2	0	72	69	56	90	1,824	109	16,700	1,010	90,306
30	0	0	72	70	154	102	1,876	110	17,906	550	107,662
31	0	0	72	71	322	90	2,108	111	16,702	314	124,050
32	0	2	70	72	532	60	2,580	112	13,552	384	137,218
33	0	4	66	73	714	30	3,264	113	9,612	844	145,986
34	2	6	62	74	786	10	4,040	114	6,250	1,766	150,470
35	8	4	66	75	714	2	4,752	115	4,664	3,108	152,026
36	20	2	84	76	532	2	5,282	116	5,784	4,608	153,202
37	32	0	116	77	322	12	5,592	117	10,254	5,814	157,642
38	38	0	154	78	156	42	5,706	118	18,194	6,278	169,558
39	32	0	186	79	72	100	5,678	119	28,730	5,814	192,474

alive in year $2i$ being equal to $B_i + B_{i-1} + \cdots + B_{i-19}$, we also get that

$$P_i \simeq B_i \left(1 + \frac{1}{r} + \cdots + \frac{1}{r^{19}} \right) = B_i \frac{1 - r^{20}}{r^{19} - r^{20}} \simeq 9.59\, B_i \,.$$

The total population is about ten times the number of births.

The proof that the sequence (B_i) shown in Table 3.1 does indeed grow asymptotically like r^i is more complicated. It was known since the work of Abraham de Moivre on recurrent series that, by introducing the "generating function"

$$f(x) = \sum_{i=0}^{\infty} B_i x^i \,,$$

one could express $f(x)$ as a rational function. Euler had explained the method in his *Introduction to Analysis of the Infinite* in 1748: the recurrent relation (3.5) gives indeed

$$f(x) = \sum_{i=0}^{12} B_i x^i + \sum_{i=13}^{\infty} (B_{i-11} + B_{i-12} + B_{i-13}) x^i$$
$$= 2x + 2x^2 + 2x^3 + 2x^{12} + f(x)(x^{11} + x^{12} + x^{13}) \,.$$

So

$$f(x) = \frac{2x + 2x^2 + 2x^3 + 2x^{12}}{1 - x^{11} - x^{12} - x^{13}} \,.$$

Euler knew that such a rational function could be decomposed in the form

$$f(x) = \frac{a_1}{1 - \frac{x}{x_1}} + \cdots + \frac{a_{13}}{1 - \frac{x}{x_{13}}} \,,$$

the numbers $x_1, \ldots, x_{13}$ being the real or complex roots of the equation $1 - x^{11} - x^{12} - x^{13} = 0$. So

$$f(x) = \sum_{i \geq 0} a_1 \left(\frac{x}{x_1} \right)^i + \cdots + a_{13} \left(\frac{x}{x_{13}} \right)^i \,.$$

Since B_i is the coefficient of x^i in $f(x)$, Euler obtained that

$$B_i = \frac{a_1}{x_1^i} + \cdots + \frac{a_{13}}{x_{13}^i} \simeq \frac{a_k}{x_k^i}$$

as $i \to +\infty$, where x_k is the root with the smallest modulus. In other words, B_i tends to grow geometrically like $(1/x_k)^i$. It remained to note that x_k is a root of the equation $1 - x^{11} - x^{12} - x^{13} = 0$ if and only if $r = 1/x_k$ is a root of equation (3.6). Certain details of the proof were finally clarified by Gumbel in 1916.

Süssmilch published a third edition of his treatise in 1765 and died in Berlin in 1767. On bad terms with the king of Prussia, Euler returned to Saint Petersburg in 1766. Despite losing his sight, he continued to publish a great number of works with the help of his sons and colleagues, especially on algebra, integral calculus, optics and shipbuilding. His *Letters on Different Subjects in Natural Philosophy Addressed to a German Princess*, written in Berlin between 1760 and 1762, was published between 1768 and 1772 and became a bestseller throughout Europe. Euler died in Saint Petersburg in 1783. His contribution to mathematical demography, especially his analysis of the "stable" age pyramid in an exponentially growing population, would be rediscovered only in the twentieth century (see Chapters 10 and 21).

Further reading

1. Euler, L.: Recherches générales sur la mortalité et la multiplication du genre humain. *Hist. Acad. R. Sci. B.-Lett. Berl.* **16**, 144–164 (1760/1767). math. dartmouth.edu/~euler/
2. Euler, L.: Sur la multiplication du genre humain. In: *Leonhardi Euleri Opera omnia*, Ser. I, vol. 7, pp. 545–552. Teubner, Leipzig (1923)
3. Euler, L.: *Introductio in analysin infinitorum*, Tomus primus. Bousquet, Lausanne (1748). Also in: Leonhardi Euleri *Opera omnia*, Ser. I, vol. 8, Teubner, Leizig (1922). English translation, Springer, New York (1988). gallica. bnf.fr
4. Fellmann, E.A.: *Leonhard Euler*. Birkhäuser, Basel (2007)
5. Gumbel, E.J.: Eine Darstellung statistischer Reihen durch Euler. *Jahresber. dtsch. Math. Ver.* **25**, 251–264 (1917). digizeitschriften.de
6. Reimer, K.F.: Johann Peter Süssmilch, seine Abstammung und Biographie. *Arch. soz. Hyg. Demogr.* **7**, 20–28 (1932)
7. Rohrbasser, J.M.: Johann Peter Süssmilch. In: Heyde, C.C., Seneta, E. (eds.) *Statisticians of the Centuries*, pp. 72–76. Springer, New York (2001)
8. Süssmilch, J.P.: *Die göttliche Ordnung in den Veränderungen des menschlichen Geschlechts aus der Geburt, dem Tode und der Fortpflanzung desselben*. Berlin (1761). echo.mpiwg-berlin.mpg.de
9. Warusfel, A.: *Euler, les mathématiques et la vie*. Vuibert, Paris (2009)

Chapter 4
Daniel Bernoulli, d'Alembert
and the inoculation of smallpox (1760)

Daniel Bernoulli was born in 1700 in Groningen in the Netherlands. His family included already two famous mathematicians: his father Johann Bernoulli and his uncle Jakob Bernoulli. In 1705 Johann moved to Basel in Switzerland where he took the professorship left vacant by the death of Jakob. Johann did not want his son to study mathematics. So Daniel turned to medicine, obtaining his doctoral degree in 1721 with a thesis on respiration. He moved to Venice and began focusing on mathematics, publishing a book in 1724. Having won a prize from the Paris Academy of Sciences that same year for an essay "On the perfection of the hourglass on a ship at sea", he obtained a professorship at the new Saint Petersburg Academy. During these years, he worked especially on recurrent series or on the "paradox of Saint Petersburg" in probability theory. In 1733 Daniel Bernoulli returned to the University of Basel, where he taught successively botany, physiology and physics. In 1738 he published a book on fluid dynamics that has remained famous in the history of physics. Around 1753 he became interested at the same time as Euler and d'Alembert in the problem of vibrating strings, which caused an important mathematical controversy.

Fig. 4.1 Daniel Bernoulli
(1700–1782)

N. Bacaër, *A Short History of Mathematical Population Dynamics*,
DOI 10.1007/978-0-85729-115-8_4, © Springer-Verlag London Limited 2011

In 1760 he submitted to the Academy of Sciences in Paris a work entitled *An attempt at a new analysis of the mortality caused by smallpox and of the advantages of inoculation to prevent it*. The question was whether inoculation (the voluntary introduction of a small amount of less virulent smallpox in the body to protect it against later infections) should be encouraged even if it is sometimes a deadly operation. This technique had been known for a long time in Asia and had been introduced in 1718 in England by Lady Montagu, wife of the British ambassador to the Ottoman Empire. In France, despite the death of the eldest son of Louis XIV from smallpox in 1711, inoculation was considered reluctantly. Voltaire, who had survived from smallpox in 1723 and who had lived several years in exile in England observing the latest innovations, pleaded for inoculation in his *Philosophical Letters* in 1734. The French scientist La Condamine, who had also survived from smallpox, pleaded for inoculation at the Academy of Sciences in Paris in 1754.

Before dying in Basel in 1759, Maupertuis encouraged Daniel Bernoulli to study the inoculation problem from a mathematical point of view. More precisely, the challenge was to find a way of comparing the long-term benefit of inoculation with the immediate risk of dying. For this purpose, Bernoulli made the following simplifying assumptions:

- people infected with smallpox for the first time die with a probability p (independent of age) and survive with a probability $1 - p$;
- everybody has a probability q of being infected each year; more precisely, the probability for one individual to become infected between age x and age $x + dx$ is $q\,dx$, where dx is an infinitesimal time period;
- people surviving from smallpox are protected against new infections for the rest of their life (they have been immunized).

Let $m(x)$ be the mortality at age x due to causes other than smallpox: the probability for one individual to die in an infinitesimal time period dx between age x and age $x + dx$ is $m(x)\,dx$. Considering a group of P_0 people born the same year, let us call

- $S(x)$ the number of "susceptible" people[1] who are still alive at age x without ever having been infected with smallpox;
- $R(x)$ the number of people who are alive at age x and who have survived from smallpox;
- $P(x) = S(x) + R(x)$ the total number of people alive at age x.

Birth corresponds to age $x = 0$. So $S(0) = P(0) = P_0$ and $R(0) = 0$. Applying the methods of calculus that had been developed at the end of the seventeenth century by Newton, Leibniz and later by his father, Daniel Bernoulli noticed that, between age x and age $x + dx$ (with dx infinitely small), each susceptible individual has a probability $q\,dx$ of being infected with smallpox and a probability $m(x)\,dx$ of dying from other causes. So the variation of the number of susceptible people is $dS = -Sq\,dx - Sm(x)\,dx$, leading to the differential equation

---

[1] More exactly, it is the expectation of this number, which can vary continuously and not just by units of one.

$$\frac{dS}{dx} = -qS - m(x)S. \tag{4.1}$$

In this equation, dS/dx is called the derivative of the function $S(x)$. During the same small time interval, the number of people dying from smallpox is $pSqdx$ and the number of people who survive from smallpox is $(1-p)Sqdx$. Moreover, there are also $Rm(x)dx$ people who die from causes other than smallpox. This leads to a second differential equation:

$$\frac{dR}{dx} = q(1-p)S - m(x)R. \tag{4.2}$$

Adding the two equations, we get

$$\frac{dP}{dx} = -pqS - m(x)P. \tag{4.3}$$

From equations (4.1) and (4.3), Bernoulli could show that the fraction of people who are still susceptible at age x is

$$\frac{S(x)}{P(x)} = \frac{1}{(1-p)e^{qx}+p}. \tag{4.4}$$

To get formula (4.4), Bernoulli eliminated $m(x)$ from the equations (4.1) and (4.3):

$$-m(x) = q + \frac{1}{S}\frac{dS}{dx} = pq\frac{S}{P} + \frac{1}{P}\frac{dP}{dx}.$$

It follows after rearrangement that

$$\frac{1}{P}\frac{dS}{dx} - \frac{S}{P^2}\frac{dP}{dx} = -q\frac{S}{P} + pq\left[\frac{S}{P}\right]^2.$$

We notice that the left-hand-side is the derivative of $f(x) = S(x)/P(x)$, which is the fraction of susceptible people in the population aged x. So

$$\frac{df}{dx} = -qf + pqf^2. \tag{4.5}$$

The solution of this type of equation had been known for several decades thanks to the work of Jakob Bernoulli, Daniel's uncle. Dividing the equation by f^2 and setting $g(x) = 1/f(x)$, we see that $dg/dx = qg - pq$ and that $g(0) = 1/f(0) = 1$. Setting $h(x) = g(x) - p$, we get $dh/dx = qh$. So $h(x) = h(0)e^{qx} = (1-p)e^{qx}$. Finally $g(x) = (1-p)e^{qx} + p$ and $f(x) = 1/g(x)$. Q.E.D.

To apply his theory, Bernoulli used Halley's life table (see Chapter 2). This table gives the number of people still alive at the beginning of year x (with $x = 1,2\ldots$) among a cohort of 1,238 born during year 0. But in the framework of his model,

Bernoulli needed the number of people $P(x)$ that actually reach age x, which is slightly different. Because Bernoulli – like most of his contemporaries – did not realize the difference (Halley's article is not very explicit indeed), he kept the numbers in Halley's table except the first number $1,238$, which he replaced by $1,300$ to get a realistic mortality during the first year of life. These numbers are shown in the second column of Table 4.1.

Table 4.1 Halley's life table and Bernoulli's computations.

Age x	Alive $P(x)$	Susceptible $S(x)$	Immune $R(x)$	Smallpox deaths	No smallpox $P^*(x)$
0	1300	1300	0	17.2	1300
1	1000	896	104	12.3	1015
2	855	685	170	9.8	879
3	798	571	227	8.2	830
4	760	485	275	7.0	799
5	732	416	316	6.1	777
6	710	359	351	5.2	760
7	692	311	381	4.6	746
8	680	272	408	4.0	738
9	670	238	432	3.5	732
10	661	208	453	3.0	726
11	653	182	471	2.7	720
12	646	160	486	2.3	715
13	640	140	500	2.1	711
14	634	123	511	1.8	707
15	628	108	520	1.6	702
16	622	94	528	1.4	697
17	616	83	533	1.2	692
18	610	72	538	1.1	687
19	604	63	541	0.9	681
20	598	55	543	0.8	676
21	592	49	543	0.7	670
22	586	42	544	0.6	664
23	579	37	542	0.5	656
24	572	32	540		649
⋮	⋮	⋮	⋮	⋮	⋮

Bernoulli chose for the probability of dying from smallpox $p = 1/8 = 12.5\%$, which is in agreement with the observations of his time. The annual probability of catching smallpox q could not be estimated directly. So Bernoulli probably tried several values for q and finally chose the one such that the number of deaths due to smallpox after all the computations below is about $1/13$ of the total number of deaths, a proportion which had then been observed in several European cities. The choice $q = 1/8$ per year turned out to give a good fit, as we shall now see[2].

With formula (4.4) and the values of $P(x)$ in the second column of the table, we can compute the number $S(x)$ of susceptible people aged x: this is the third column of the table rounded to the nearest integer. The fourth column shows the number $R(x) = P(x) - S(x)$ of people aged x having survived from smallpox. The fifth column shows in the row corresponding to age x the number of deaths due to smallpox between age x and age $x+1$. In theory this number should be the integral $pq \int_x^{x+1} S(t)\,dt$ but the formula $pq[S(x) + S(x+1)]/2$ gives a good approximation, as sketched in Fig. 4.2: the area of the trapezoid is close to the area under the curve, i.e. to the integral of the function.

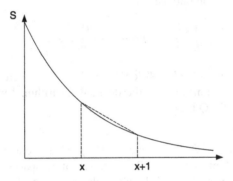

Fig. 4.2 The area of the dashed trapezoid approximates the integral of the function S between x and $x+1$.

Bernoulli noticed that the sum of all the numbers in the fifth column gives 98 deaths from smallpox before the age of 24. If we continued the table for older ages, we would find only three more deaths from smallpox among the 32 people who are still susceptible at age 24. In summary, starting from $1,300$ births, the fate of 101 people is to die from smallpox. This is almost exactly the expected fraction $1/13$.

Bernoulli considered then the situation where smallpox would be inoculated to everybody at birth and would not cause any deaths. Smallpox would be eradicated and the question is to estimate the increase in life expectancy. Starting from the same number of births P_0, let us call $P^*(x)$ the number of people aged x when smallpox has disappeared. Then

$$\frac{dP^*}{dx} = -m(x)P^*. \tag{4.6}$$

Bernoulli could show that

---

[2] The fact that p and q are equal is just a coincidence.

$$P^*(x) = \frac{P(x)}{1 - p + pe^{-qx}}, \tag{4.7}$$

where $P(x)$ is as above the population aged x when smallpox is present.

Indeed, eliminating as before $m(x)$ between equations (4.6) and (4.3), Bernoulli obtained after rearrangement

$$\frac{1}{P^*}\frac{dP}{dx} - \frac{P}{P^{*2}}\frac{dP^*}{dx} = -pq\frac{S}{P}\frac{P}{P^*}.$$

He set $h(x) = P(x)/P^*(x)$. Using formula (4.4), he multiplied numerator and denominator by e^{-qx} and obtained

$$\frac{1}{h}\frac{dh}{dx} = -pq\frac{e^{-qx}}{1 - p + pe^{-qx}},$$

which is equivalent to $\frac{d}{dx}\log h = \frac{d}{dx}\log(1 - p + pe^{-qx})$, where log stands here for the natural logarithm and not the decimal logarithm. But $h(0) = 1$. So $h(x) = 1 - p + pe^{-qx}$. Q.E.D.

Notice that the ratio $P(x)/P^*(x)$ tends to $1 - p$ when the age x is high enough. The sixth column of Table 4.1 shows $P^*(x)$. A way of comparing $P(x)$ and $P^*(x)$ is to estimate the life expectancy at birth, whose theoretical expression with smallpox is

$$\frac{1}{P_0}\int_0^\infty P(x)\,dx.$$

A similar expression with $P^*(x)$ replacing $P(x)$ holds without smallpox. Bernoulli used the approximate formula $[\frac{1}{2}P(0) + P(1) + P(2) + \cdots]/P_0$, which is the one given by the method of trapezoids (Fig. 4.2). Continuing the table beyond 24 years until 84 years (see Table 2.1), he obtained finally a life expectancy E with smallpox equal to $[\frac{1}{2}1300 + 1000 + \cdots + 20]/1300 \simeq 26.57$ years, i.e. 26 years and 7 months. Without smallpox, he obtained a life expectancy E^* equal to $[\frac{1}{2}1300 + 1015 + \cdots + 23]/1300 \simeq 29.65$ years, i.e. 29 years and 8 months. Inoculation at birth would increase life expectancy by more than three years.

We can note that there is a simpler and faster method than the one used by Bernoulli to get these formulas. Starting from the differential equation (4.1) for $S(x)$, we see first that

$$S(x) = P_0 e^{-qx} \exp\left(-\int_0^x m(y)\,dy\right).$$

Using this expression in equation (4.2) for $R(x)$, we find that

$$R(x) = P_0 (1 - p)(1 - e^{-qx}) \exp\left(-\int_0^x m(y)\,dy\right).$$

Equation (4.6) for $P^*(x)$ shows that

$$P^*(x) = P_0 \exp\left(-\int_0^x m(y)\,dy\right). \tag{4.8}$$

Formulas (4.4) and (4.7) follow immediately!

Of course, inoculation with a less virulent strain of smallpox is not completely safe. If p' is the probability of dying from smallpox just after inoculation ($p' < p$), then the life expectancy would be $(1 - p')E^*$ if everybody went through inoculation at birth. This life expectancy remains higher than the "natural" life expectancy E if $p' < 1 - E/E^*$ or about 11%. Data concerning p' was difficult to obtain at the time. But Bernoulli estimated that the risk p' was less than 1%. For him there was no doubt: inoculation had to to be promoted by the State. He concluded:

> I simply wish that, in a matter which so closely concerns the wellbeing of the human race, no decision shall be made without all the knowledge which a little analysis and calculation can provide.

Bernoulli's work was presented at the Academy of Sciences in Paris in April 1760. In November, d'Alembert presented a comment entitled *On the application of probability theory to the inoculation of smallpox*. The comment was published shortly after in the second volume of his *Opuscules mathématiques* with more detailed computations and together with another work entitled *Mathematical theory of inoculation*. D'Alembert criticized Bernoulli's assumptions about the probability of infection and the probability of dying from smallpox being independent of age. He suggested a different solution that does not require these assumptions. Call $v(x)$ the mortality due to smallpox at age x, $m(x)$ the mortality due to other causes and $P(x)$ the number of people that are still alive. Then

$$\frac{dP}{dx} = -v(x)P - m(x)P. \tag{4.9}$$

Comparing with equation (4.3), we see that in fact $v(x) = pqS(x)/P(x)$. Here we obtain

$$P^*(x) = P(x) \exp\left(\int_0^x v(y)\,dy\right), \tag{4.10}$$

where $P^*(x)$ stands for the number of people alive at age x when smallpox has disappeared.

Indeed we can either substitute the function $m(x)$ between equations (4.6) and (4.9) or use formula (4.8) for $P^*(x)$ and notice that the solution of (4.9) is given by

Fig. 4.3 D'Alembert (1717–1783)

$$P(x) = P_0 \exp\left(-\int_0^x [v(y) + m(y)] \, dy\right).$$

Formula (4.10) given by d'Alembert does not contradict Bernoulli's formula (4.7). It just uses a different type of information $v(x)$, which was not available at the time because death registers included the cause of death but not the age of the victim. D'Alembert suggested that one could not really conclude whether inoculation was useful before this type of data became available.

D'Alembert also criticized the usefulness of the life expectancy as a criterion for decision since it gives the same weight to all the years, whether in a near or distant future. He noticed that, from the point of view of the individual or of the State, not all the years have the same "utility", young and old ages being less valuable than mean ages. Despite all these criticisms, d'Alembert declared himself in favor of inoculation.

Because of publication delays, Bernoulli's work was published only in 1766, while d'Alembert managed to get his own work published very quickly. Bernoulli expressed his bitterness in a letter to Euler:

What do you say about the enormous platitudes of the great d'Alembert about probabilities: as I find myself too frequently unjustly treated in his publications, I have decided already some time ago to read nothing anymore which comes from his pen. I have taken this decision on the occasion of a manuscript about inoculation which I sent to the Academy in Paris eight years ago and which was greatly appreciated because of the novelty of the analysis. It was, I dare say, like incorporating a new province into the body of mathematics. It seems that the success of this new analysis caused him pains of the heart. He has criticized it in a thousand ways all equally ridiculous and after having it well criticized, he pretends to be the first author of a theory which he did not only hear mentioned. He, however, knew that my manuscript could only appear after some seven or eight years. He could only have knowledge about it in his capacity as member of the Academy. In this respect my manuscript should have stayed sacred until it was made public. *Dolus an virtus quis in hoste requirat!*[3]

---

[3] *What matters whether by valor or by stratagem we overcome the enemy!* Vergil: Aeneid, Book II.

Despite the works of Bernoulli and d'Alembert, inoculation was not performed on a large scale in France. King Louis XV died of smallpox in 1774. The medical doctors of the court did inoculate the rest of the royal family shortly after. The problem lost its importance when Edward Jenner discovered that inoculating cowpox to humans ("vaccination") protected against smallpox and was safe. His work, *An inquiry into the causes and effects of the variolae vaccina*, was published in 1798. Vaccination spread rapidly throughout Europe. Nevertheless the methods developed for the computation of the increase in life expectancy if one cause of death is removed are still in use today.

In the following decades, data concerning the age at which people died of smallpox became available. The problem was reconsidered especially by

- Johann Heinrich Lambert, a mathematician from the Berlin Academy, in 1772;
- Emmanuel-Étienne Duvillard, then in charge of population statistics at the Interior Ministry in Paris, in his *Analysis and Tables of the Influence of Smallpox on the Mortality at each Age* (1806);
- Pierre-Simon Laplace in his *Analytic Theory of Probability* (1812).

Duvillard and Laplace showed for example how to modify formula (4.7) when the parameters p and q depend on age:

$$P^*(x) = \frac{P(x)}{1 - \int_0^x p(y)\, q(y)\, e^{-\int_0^y q(z)\, dz}\, dy}.$$

Here, $p(x)$ is the probability of dying of smallpox if infected at age x and $q(x)$ is the probability of being infected with smallpox at age x.

After this work on smallpox, Daniel Bernoulli did not consider any other problem in population dynamics. He died in Basel in 1782. D'Alembert died in Paris a year later.

Further reading

1. d'Alembert, J.: Onzième mémoire, Sur l'application du calcul des probabilités à l'inoculation de la petite vérole. In: *Opuscules mathématiques*, Tome second, pp. 26–95. David, Paris (1761). books.google.com
2. Bernoulli, D.: Réflexions sur les avantages de l'inoculation. *Mercure de France*, 173–190 (1760). Also in: *Die Werke von Daniel Bernoulli*, Band 2, pp. 268–274. Birkhäuser, Basel (1982). books.google.com
3. Bernoulli, D.: Essai d'une nouvelle analyse de la mortalité causée par la petite vérole et des avantages de l'inoculation pour la prévenir. *Hist. Acad. R. Sci. Paris*, 1–45 (1760/1766). English translation: *Rev. Med. Virol.* **14**, 275–288 (2004). gallica.bnf.fr
4. Dietz, K., Heesterbeek, J.A.P.: Daniel Bernoulli's epidemiological model revisited. *Math. Biosci.* **180**, 1–21 (2002)

5. Duvillard, E.E.: *Analyse et tableaux de l'influence de la petite vérole sur la mortalité à chaque âge*. Imprimerie Impériale, Paris (1806). `books.google.com`

6. Lambert, J.H.: *Contributions mathématiques à l'étude de la mortalité et de la nuptialité* (1765 et 1772). INED, Paris (2006). `books.google.com`

7. Laplace, P.S.: *Théorie analytique des probabilités*. Courcier, Paris (1812). `books.google.com`

8. Straub, H.: Bernoulli, Daniel. In: Gillespie, C.C. (ed.) *Dictionary of Scientific Biography*, vol. 2, pp. 36–46. Scribner, New York (1970)

9. Tent, M.B.W.: *Leonhard Euler and the Bernoullis*. A K Peters, Natick, Massachusetts (2009)

10. Voltaire: *Lettres philosophiques*. Lucas, Amsterdam (1734). `gallica.bnf.fr`

Chapter 5
Malthus and the obstacles to geometric growth (1798)

Thomas Robert Malthus was born in 1766 near London, the sixth of seven children. His father, a friend and admirer of Jean-Jacques Rousseau, was his first teacher. In 1784 the young Malthus started studying mathematics at Cambridge University. He obtained his diploma in 1791, became a fellow of Jesus College in 1793 and an Anglican priest in 1797.

Fig. 5.1 Malthus
(1766–1834)

In 1798 Malthus published anonymously a book entitled *An Essay on the Principle of Population, as It Affects the Future Improvement of Society, With Remarks on the Speculations of Mr Godwin, Mr Condorcet and Other Writers*. It came as a reaction against Godwin's *Enquiry Concerning Political Justice* (1793) and Condorcet's *Sketch for a Historical Picture of the Progress of the Human Mind* (1794). Despite the horrors that the French Revolution did in the name of progress, the two authors claimed that the progress of society was inevitable. Malthus did not share the same optimism. He also argued that the English Poor Laws, which helped poor families with many children, favored the growth of the population without encouraging a similar growth in the production of food. It seemed to him that these laws did not really relieve the poor; quite the contrary. More generally, population tend-ing to grow always faster than the production of food, part of society seemed to be condemned to misery, hunger or epidemics: these are the scourges that slow down population growth and that, in Malthus' opinion, are the principal obstacles to the progress of society. All the theories promising progress would just be utopian. These ideas led Malthus to publish his book in 1798. Here is how he summarized his thesis:

> [...] the power of population is indefinitely greater than the power in the earth to produce subsistence for man. Population, when unchecked, increases in a geometrical ratio. Subsistence increases only in an arithmetical ratio. A slight acquaintance with numbers will shew the immensity of the first power in comparison of the second. By that law of our nature which makes food necessary to the life of man, the effects of these two unequal powers must be kept equal. This implies a strong and constantly operating check on population from the difficulty of subsistence. This difficulty must fall somewhere; and must necessarily be severely felt by a large portion of mankind.

Malthus' book was very successful. It contained few data. Malthus noticed, for example, that the population of the USA had doubled every twenty five years during the eighteenth century. He did not really try to translate his theses into mathematical models but paved the way for later work by Adolphe Quetelet and Pierre-François Verhulst, who will be the subject of the next chapter.

After the publication of his book, Malthus traveled with friends first to Germany, Scandinavia and Russia, then to France and Switzerland. Putting together the information collected during his journeys, he published under his name a very much enlarged second edition in 1803, with a different subtitle: *An Essay on the Principle of Population, or a View of its Past and Present Effects on Human Happiness, With an Enquiry Into Our Prospects Respecting the Future Removal or Mitigation of the Evils Which It Occasions*. This new edition discussed in detail the obstacles to population growth in various countries: delayed marriage, abortion, infanticide, famine, war, epidemics, economic factors.... For Malthus, delayed marriage was the best option to stabilize the population. Four other editions of the book followed in 1806, 1807, 1817 and 1826. In 1805 Malthus became professor of history and political economy in a new school set up by the West Indies Company for its employees. He also published *An Inquiry into the Nature and Progress of Rent* (1815) and *Principles of Political Economy* (1820). In 1819 Malthus was elected to the Royal Society. In 1834 he was one the founding members of the Statistical Society. He died near Bath that same year.

Malthus' work had a strong influence on the development of the theory of evolution. Charles Darwin, back from his journey on board the *Beagle*, read Malthus' book on population in 1838. Here is what he wrote in the introduction to his famous book *On the Origin of Species by Means of Natural Selection*, published in 1859:

> In the next chapter the Struggle for Existence amongst all organic beings throughout the world, which inevitably follows from their high geometrical powers of increase, will be treated of. This is the doctrine of Malthus, applied to the whole animal and vegetable kingdoms.

Alfred Russel Wallace, who developed the theory of evolution at the same time as Darwin, also said that his ideas came after reading Malthus' book.

In contrast here is the point of view of Karl Marx on the success of Malthus' book, as can be read in a footnote of his *Capital*:

> If the reader reminds me of Malthus, whose *Essay on Population* appeared in 1798, I remind him that this work in its first form is nothing more than a schoolboyish, superficial plagiary of De Foe, Sir James Steuart, Townsend, Franklin, Wallace, etc., and does not contain a single sentence thought out by himself. The great sensation this pamphlet caused, was due solely to party interest. The French Revolution had found passionate defenders in the United Kingdom; the *principle of population*, slowly worked out in the eighteenth century, and then, in the midst of a great social crisis, proclaimed with drums and trumpets as the infallible antidote to the teachings of Condorcet, etc., was greeted with jubilance by the English oligarchy as the great destroyer of all hankerings after human development. Malthus, hugely astonished at his success, gave himself to stuffing into his book materials superficially compiled and adding to it new matter not discovered but annexed by him.

Certainly Malthus' theses were not completely new. For example, the idea that population tends to grow geometrically is often attributed[1] to him, even though we saw in Chapter 3 that this idea was already familiar to Euler half a century earlier. However, Malthus gave it publicity by linking it in a polemic way to real legislative problems. Ironically it was in communist China that Malthus' suggestion to limit births would find its most striking application (see Chapter 25).

Further reading

1. Condorcet: *Esquisse d'un tableau historique des progrès de l'esprit humain*. Agasse, Paris (1794). `gallica.bnf.fr`
2. Darwin, C.: *On the Origin of Species by Means of Natural Selection, or the Preservation of Favoured Races in the Struggle for Life*. John Murray, London (1859). `darwin-online.org.uk`
3. Godwin, W.: *An Enquiry Concerning Political Justice*. Robinson, London (1793). `www.archive.org`
4. Malthus, T.R.: *An Essay on the Principle of Population*, 1st edn. London (1798). `www.econlib.org`
5. Marx, K.: *Capital, A Critical Analysis of Capitalist Production*, vol. 1. London (1887). `www.archive.org`
6. Simpkins, D.M.: Malthus, Thomas Robert. In: Gillespie, C.C. (ed.) *Dictionary of Scientific Biography*, vol. 9, pp. 67–71. Scribner, New York (1974)

---

[1] R. A. Fisher (see Chapters 14 and 20) would call "Malthusian parameter" the growth rate of populations. Malthus did mention the treatise of Süssmilch in his own book.

Chapter 6
Verhulst and the logistic equation (1838)

Pierre-François Verhulst was born in 1804 in Brussels. He obtained a PhD in mathematics from the University of Ghent in 1825. He was also interested in politics. While in Italy to contain his tuberculosis, he pleaded without success in favor of a constitution for the Papal States. After the revolution of 1830 and the independence of Belgium, he published a historical essay on an eighteenth century patriot. In 1835 he became professor of mathematics at the newly created Free University in Brussels.

Fig. 6.1 Verhulst
(1804–1849)

That same year 1835, his compatriot Adolphe Quetelet, a statistician and director of the observatory in Brussels, published *A Treatise on Man and the Development of his Faculties*. Quetelet suggested that populations could not grow geometrically over a long period of time because the obstacles mentioned by Malthus formed a kind of "resistance", which he thought (by analogy with mechanics) was proportional to the square of the speed of population growth. This analogy had no real basis, but it inspired Verhulst.

N. Bacaër, *A Short History of Mathematical Population Dynamics*, 35
DOI 10.1007/978-0-85729-115-8_6, © Springer-Verlag London Limited 2011

Indeed, Verhulst published in 1838 a *Note on the law of population growth*. Here are some extracts:

> We know that the famous Malthus showed the principle that the human population tends to grow in a geometric progression so as to double after a certain period of time, for example every twenty five years. This proposition is beyond dispute if abstraction is made of the increasing difficulty to find food [...]
>
> The virtual increase of the population is therefore limited by the size and the fertility of the country. As a result the population gets closer and closer to a steady state.

Verhulst probably realized that Quetelet's mechanical analogy was not reasonable and proposed instead the following (still somewhat arbitrary) differential equation for the population $P(t)$ at time t:

$$\frac{dP}{dt} = rP\left(1 - \frac{P}{K}\right). \tag{6.1}$$

When the population $P(t)$ is small compared to the parameter K, we get the approximate equation

$$\frac{dP}{dt} \simeq rP,$$

whose solution is $P(t) \simeq P(0)\,e^{rt}$, i.e. exponential growth[1]. The growth rate decreases as $P(t)$ gets closer to K. It would even become negative if $P(t)$ could exceed K. To get the exact expression of the solution of equation (6.1), we can proceed like Daniel Bernoulli for equation (4.5).

Dividing equation (6.1) by P^2 and setting $p = 1/P$, we get $dp/dt = -rp + r/K$. With $q = p - 1/K$, we get $dq/dt = -rq$ and $q(t) = q(0)\,e^{-rt} = (1/P(0) - 1/K)\,e^{-rt}$. So we can deduce $p(t)$ and $P(t)$.

Finally we get after rearrangement

$$P(t) = \frac{P(0)\,e^{rt}}{1 + P(0)\,(e^{rt} - 1)/K}. \tag{6.2}$$

The total population increases progressively from $P(0)$ at time $t = 0$ to the limit K, which is reached only when $t \to +\infty$ (Fig. 6.2). Without giving the values he used for the unknown parameters r and K, Verhulst compared his result with data concerning the population of France between 1817 and 1831, of Belgium between 1815 and 1833, of the county of Essex in England between 1811 and 1831, and of Russia between 1796 and 1827. The fit turned out to be pretty good.

In 1840 Verhulst became professor at the Royal Military School in Brussels. The following year he published an *Elementary Treatise of Elliptic Functions* and was elected to the Royal Academy of Belgium. In 1845 he continued his population

---

[1] One usually speaks of geometric growth in discrete-time models and of exponential growth in continuous-time models but is is essentially the same thing.

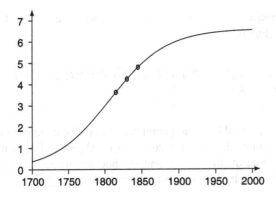

Fig. 6.2 The population of Belgium (in millions) and the logistic curve. The data points correspond to the years 1815, 1830 and 1845. The parameter values are those of the article from 1845.

studies with an article entitled "Mathematical enquiries on the law of population growth". He first turned back to Malthus' remark according to which the population of the USA had doubled every 25 years (Tab. 6.1). If we compute the ratio between

Table 6.1 Official censuses of the population of the USA.

Year	Population
1790	3,929,827
1800	5,305,925
1810	7,239,814
1820	9,638,131
1830	12,866,020
1840	17,062,566

the population in year $n + 10$ to that in year n, we find respectively 1.350, 1.364, 1.331, 1.335 and 1.326, which is fairly constant. The population was hence multiplied on average by 1.34 every 10 years and by $1.34^{25/10} \simeq 2.08$ every 25 years. So it had continued to double every 25 years since Malthus' essay, almost half a century earlier. However Verhulst added:

> We shall not insist on the hypothesis of geometric progression, given that it can hold only in very special circumstances; for example, when a fertile territory of almost unlimited size happens to be inhabited by people with an advanced civilization, as was the case for the first American colonies.

In his article Verhulst also returned to equation (6.1), which he called "logistic". He noticed that the curve $P(t)$ increases with a positive curvature (it is convex) as long as $P(t) < K/2$ and then continues to increase towards K but with a negative curvature (it is concave) as soon as $P(t) > K/2$. So the curve has the shape of a distorted letter S (Fig. 6.2).

> Indeed, $d^2P/dt^2 = r(1 - 2P/K)dP/dt$. So $d^2P/dt^2 > 0$ if $P < K/2$ and $d^2P/dt^2 < 0$ if $P > K/2$.

Verhulst also explained how the parameters r and K can be estimated from the population $P(t)$ in three different but equally spaced years. If P_0 is the population at time $t = 0$, P_1 that at time $t = T$ and P_2 that at time $t = 2T$, then a tedious computation starting from equation (6.2) shows that

$$K = P_1 \frac{P_0 P_1 + P_1 P_2 - 2 P_0 P_2}{P_1^2 - P_0 P_2}, \quad r = \frac{1}{T} \log\left[\frac{1/P_0 - 1/K}{1/P_1 - 1/K}\right].$$

Using the estimations for the population of Belgium in the years 1815, 1830 and 1845 (respectively 3.627, 4.247 and 4.801 million), he obtained $K = 6.584$ million and $r = 2.62\%$ per year. He could then use equation (6.2) to predict that the population of Belgium would be 4.998 million at the beginning of the year 1851 and 6.064 million at the beginning of the year 1900 (Fig. 6.2). Verhulst did a similar study for France. He obtained $K = 39.685$ million and $r = 3.2\%$ per year. As the populations of Belgium and France have in the mean time largely exceeded these values of K, we see that the logistic equation can be a realistic model only for periods of time of a few decades, as in Verhulst's 1838 article, but not for longer periods.

In 1847 appeared a *Second enquiry on the law of population growth* in which Verhulst gave up the logistic equation and chose instead a differential equation that can be written in the form

$$\frac{dP}{dt} = r\left(1 - \frac{P}{K}\right).$$

He thought that this equation would hold when the population $P(t)$ is above a certain threshold. The solution is

$$P(t) = K + (P(0) - K)e^{-rt/K}.$$

Using the same demographic data for Belgium, Verhulst estimated anew the parameters r and K. This time he found $K = 9.4$ million for the maximum population. We see how much the result can depend on the choice of the model!

Verhulst became president of the Royal Academy of Belgium in 1848, but died the following year in Brussels, probably of tuberculosis. Despite Verhulst's hesitation between model equations, the logistic equation was reintroduced independently several decades later by different people. Robertson used it in 1908 to model the individual growth of animals, plants, humans and body organs. McKendrick and Kesava Pai used it in 1911 for the growth of populations of microorganisms. Pearl and Reed used it in 1920 for the growth of the population of the USA, which had started to slow down. In 1922 Pearl finally noticed the work of Verhulst. From then on, the logistic equation inspired many works (see Chapters 13, 20 and 24). The maximum population K eventually became known as the "carrying capacity".

Further reading

1. Lloyd, P.J.: American, German and British antecedents to Pearl and Reed's logistic curve. *Pop. Stud.* **21**, 99–108 (1967)
2. McKendrick, A.G., Kesava Pai, M.: The rate of multiplication of microorganisms: A mathematical study. *Proc. R. Soc. Edinb.* **31**, 649–655 (1911)
3. Pearl, R.: *The Biology of Death.* Lippincott, Philadelphia (1922). www. archive.org
4. Pearl, R., Reed, L.J.: On the rate of growth of the population of the United States since 1790 and its mathematical representation. *Proc. Natl. Acad. Sci.* **6**, 275–288 (1920). www.pnas.org
5. Quetelet, A.: *Sur l'homme et le développement de ses facultés.* Bachelier, Paris (1835). gallica.bnf.fr
6. Quetelet, A.: Pierre-François Verhulst. *Annu. Acad. R. Sci. Lett. B.-Arts Belg.* **16**, 97–124 (1850). books.google.com
7. Quetelet, A.: *Sciences mathématiques et physiques au commencement du* XIX*ième siècle.* Mucquardt, Bruxelles (1867). gallica.bnf.fr
8. Robertson, T.B.: On the normal rate of growth of an individual and its biochemical significance. *Arch. Entwicklungsmechanik Org.* **25**, 581–614 (1908)
9. Verhulst, P.-F.: Notice sur la loi que la population poursuit dans son accroissement. *Corresp. Math. Phys.* **10**, 113–121 (1838). books.google.com
10. Verhulst, P.-F.: Recherches mathématiques sur la loi d'accroissement de la population. *Nouv. Mém. Acad. R. Sci. B.-lett. Brux.* **18**, 1–45 (1845). gdz.sub. uni-goettingen.de
11. Verhulst, P.-F.: Deuxième mémoire sur la loi d'accroissement de la population. *Mém. Acad. R. Sci. Lett. B.-Arts Belg.* **20**, 142–173 (1847). www.digizeit-schriften.de

Chapter 7
Bienaymé, Cournot and the extinction of family names (1845–1847)

Irenée Jules Bienaymé was born in 1796 in Paris. He studied at the École Poly-technique and made a career in the Ministry of Finances, reaching the high level of general inspector. Influenced by the book *Analytic Theory of Probability* writ-ten by Laplace, Bienaymé also found time to publish articles on many applications of probability theory, such as demographic and medical statistics (infant mortality, number of births, life expectancy), probability of errors in justice, insurance theory and representativeness of voting systems.

Fig. 7.1 Bienaymé
(1796–1878)

In 1845 Bienaymé wrote a short note "On the law of multiplication and the dura-tion of families", which was published in the bulletin of the Société Philomatique in Paris. A number of authors had already written on this subject. In the second edition of *An Essay on the Principle of Population* (1803), Malthus included a chapter on the population of Switzerland and noticed that

N. Bacaër, *A Short History of Mathematical Population Dynamics*,
DOI 10.1007/978-0-85729-115-8_7, © Springer-Verlag London Limited 2011

in the town of Berne, from the year 1583 to 1654, the sovereign council had admitted into the Bourgeoisie 487 families, of which 379 became extinct in the space of two centuries, and in 1783 only 108 of them remained.

In 1842 Thomas Doubleday claimed more generally that upper-class families from the nobility or from the bourgeoisie had a greater tendency to disappear than lower-class families. Similar ideas were put forward in France by Émile Littré in 1844 in a text of introduction to the positivist philosophy of Auguste Comte and by Benoiston de Châteauneuf – a friend of Bienaymé – who published in 1845 an essay *On the duration of noble families in France*.

It was in this context that Bienaymé tried to explain how it could be that the population of a country tends to grow geometrically while a great number of families disappear. To attack this problem he considered the simplified case where all men would have the same probabilities of having 0, 1, 2, 3, ... sons reaching adulthood. More precisely, he asked himself what was the probability for a man to have offspring carrying his name after n generations. If the mean number of sons is less than one, it is clear that this probability should tend to zero as n grows to infinity. Bienaymé noticed that the same conclusion would remain true[1] if the mean number of sons was exactly one, e.g. if there is a probability $1/2$ of having no son and a probability $1/2$ of having two sons (Fig. 7.2). But in that case the probability of having offspring in generation n tends to zero more slowly: in the example it would still be 5% after 35 generations, i.e. after eleven or twelve centuries if there are three generations per century[2]. Bienaymé noticed finally that if the mean number of sons is greater than one, the extinction of the family line is not sure: its probability can be computed by solving some algebraic equation.

Fig. 7.2 Artificial example of family tree. The ancestor is at the top of the tree. In each generation, men have a probability $1/2$ of having no son and a probability $1/2$ of having two sons.

Bienaymé's article did not contain more explanation. In 1847 his friend Antoine-Augustin Cournot, a mathematician and economist, included some details in a book

---

[1] Except if every man has exactly one son.

[2] As will shall see below, this probability is equal to $1 - x_{35}$ with $x_{n+1} = \frac{1}{2} + \frac{1}{2}x_n^2$ and $x_0 = 0$.

entitled *On the Origin and on the Limits of the Correspondence between Algebra and Geometry*. He presented the problem in the form of a game of chance but acknowledged that it was identical to Bienaymé's study of the extinction of family names. If we keep the interpretation in terms of family names, Cournot considered first the special case where men have at most two sons, p_0, p_1 and p_2 being respectively the probability of having 0, 1 or 2 sons. Of course, $p_0 + p_1 + p_2 = 1$. Starting from one ancestor, the probability of extinction after just one generation, call it x_1, is obviously equal to p_0. The probability of extinction within two generations is $x_2 = p_0 + p_1 x_1 + p_2 x_1^2$: either the family was already extinct in the first generation (probability p_0), or there was just one son in the first generation who had no male offspring (probability $p_1 x_1$), or there were two sons in the first generation and each of them had no male offspring (probability $p_2 x_1^2$). More generally, the probability of extinction within n generations is $x_n = p_0 + p_1 x_{n-1} + p_2 (x_{n-1})^2$. Indeed, if there are for example two sons in the first generation (probability p_2), the family will be extinct $n - 1$ generations later (i.e. in generation n) with a probability equal to $(x_{n-1})^2$. Cournot noticed that x_n is an increasing sequence with $x_n \leq 1$ for all n. So x_n has a limit $x_\infty \leq 1$, which is a solution of the equation $x = p_0 + p_1 x + p_2 x^2$. Using $p_1 = 1 - p_0 - p_2$, this equation is equivalent to $0 = p_2(x - 1)(x - p_0/p_2)$. So there are two roots: $x = 1$ and $x = p_0/p_2$. Three cases can be distinguished depending on the average number of sons $p_1 + 2p_2$, which is also equal to $1 - p_0 + p_2$ and which we shall call $\mathscr{R}_0$. If $\mathscr{R}_0 < 1$, then $p_0/p_2 > 1$. So $x = 1$ is the only possible value for the limit x_∞. For sure the family name will go extinct. If $\mathscr{R}_0 = 1$, both roots are equal to 1 and the conclusion is the same. If $\mathscr{R}_0 > 1$, then Cournot argued that x_∞ should be equal to the second root p_0/p_2, as the extinction probability obviously has to be 0 in the special case where $p_0 = 0$.

Cournot briefly mentioned the more general case where men can have at most m sons with probabilities p_0, $p_1, \ldots, p_m$. The conclusion depends in the same way on the value of $\mathscr{R}_0 = p_1 + 2p_2 + \cdots + mp_m$, the average number of sons, with respect to 1. The equation for x_∞, which is $x = p_0 + p_1 x + \cdots + p_m x^m$, always has the root $x = 1$. It has only one other positive root, which gives the extinction probability x_∞ when $\mathscr{R}_0 > 1$.

Unfortunately Bienaymé's article and the few pages in Cournot's book went completely unnoticed at the time. The article was only noticed in the 1970s and the book pages a further twenty years later! Meanwhile the problem and its solution had been rediscovered by others and the subject developed considerably. We shall return to that in Chapters 9, 17 and 18.

Bienaymé had to quit his job in the Ministry of Finances following the 1848 revolution. The chair of probability theory at the University of Paris, for which he was certainly the best candidate, was also given to somebody else. Nevertheless Bienaymé was able to work again for the Ministry of Finances after 1850, but he resigned in 1852. Later that year, he was elected to the Academy of Sciences where he was the specialist in the field of statistics. In 1853 he proved what some modern textbooks call the "Bienaymé–Tchebychev" inequality. In 1875 he became president of the newly created Société Mathématique de France. He died in Paris in 1878.

Chapter 8
Mendel and heredity (1865)

Johann Mendel was born in 1822 in Moravia, then part of the Austrian empire and now part of the Czech Republic. His father was a peasant. With his good results in high school and his poor health, Mendel preferred to continue studying rather than work on the family farm. But he could not afford to go to the university. So in 1843 he entered the abbey of Saint Thomas in Brünn (now Brno), where he took the name of Gregor. He studied theology, but also attended some courses on agriculture. In 1847 he was ordained a priest. He taught in a high school for a few years but failed the exam for becoming an ordinary professor. Between 1851 and 1853, thanks to the support of his hierarchy, he was nevertheless able to continue his studies at the University of Vienna where he attended courses in physics, mathematics and natural sciences. After that he returned to Brünn and taught physics in a technical school.

Fig. 8.1 Mendel (1822–1884)

Between 1856 and 1863, Mendel made a series of experiments on a great number of plants in the garden of his abbey. In 1865 he presented his results at two meetings of the Natural History Society of Brünn, of which he was a member. His work, *Experiments on Plant Hybridization*, was published in German the following year in the proceedings of the Society. Mendel explained how he had come to study the variations of peas, plants which reproduce naturally by self-fertilization and whose seeds can take different easily identifiable forms: round or wrinkled, yellow or green, etc. By crossing a plant coming from a lineage with round seeds and a plant coming from a lineage with wrinkled seeds, he noticed that he always obtained hybrids that gave round seeds. He called the character "round seeds" dominant and the character "wrinkled seeds" recessive. He showed in the same way that the character "yellow seeds" was dominant and that the character "green seeds" was recessive.

Mendel then noticed that the self-fertilization of plants grown from hybrid seeds gave in the first generation new seeds which had either the dominant or the recessive character in apparently random proportions. Moreover, he noticed that by repeating the experiment many times he obtained on average about three times more seeds with the dominant character than with the recessive character. For example, he obtained in a first experiment a total of 5,474 round seeds and 1,850 wrinkled seeds, corresponding to a ratio of 2.96 to 1. A second experiment gave a total of 6,022 yellow seeds and 2,001 green seeds, corresponding to a ratio[1] of 3.01 to 1.

Mendel also noticed that among the plants grown from the seeds of the first generation with the dominant character, those that gave by self-fertilization seeds with either the dominant or the recessive character were about twice as many as those that gave seeds with the dominant character only. For example, among the 565 plants grown from round seeds of the first generation, 372 gave both round and wrinkled seeds whereas 193 gave round seeds only; the ratio is equal to 1.93. Similarly, among 519 plants grown from yellow seeds of the first generation, 353 gave both yellow and green seeds whereas 166 gave only yellow seeds; the ratio is equal to 2.13.

To explain these results, Mendel had the brilliant idea of considering the apparent character of a seed as the result of the association of two hidden factors, each of these factors being either dominant (written A) or recessive (written a). So there are three possible combinations: AA, Aa and aa. The seeds with the factors AA or Aa have the same dominant character A. The seeds with the factors aa have the recessive character a. Mendel assumed moreover that during fertilization, the pollen grains and the ovules (the gametes) transmit only one of the two factors, each with a probability $1/2$.

Hence the crossing of pure lineages AA and aa gives hybrids that all have the factors Aa and the dominant character A. The gametes of the hybrid Aa transmit the factor A with probability $1/2$ and the factor a with the probability $1/2$. Self-fertilization of a plant grown from a hybrid seed Aa therefore gives AA with probability $1/4$, Aa with probability $1/2$ and aa with probability $1/4$, as shown in Table 8.1.

---

[1] As R. A. Fisher (see Chapter 14) later noticed, the probability of arriving at experimental results so close to the theoretical value is quite small. Mendel probably "arranged" his data. For example, in the second experiment concerning $n = 6,022 + 2,001 = 8,023$ seeds, the probability for the ratio to differ from 3 by less than 0.01 is only about 10%.

Table 8.1 Possible results of the self-fertilization of a hybrid Aa and their probabilities as a function of the factors transmitted by the male gametes (in lines) and by the female gametes (in columns).

Factor	A	a
Probability	1/2	1/2
A	AA	Aa
1/2	1/4	1/4
a	Aa	aa
1/2	1/4	1/4

Mendel noticed that the proportions $AA : Aa : aa$, which were $1 : 2 : 1$, could also be obtained by the formal computation $(A + a)^2 = AA + 2Aa + aa$. Since the seeds AA and Aa have the apparent character A while only the seeds aa have the apparent character a, there are indeed three times more seeds with the character A than with the character a. Moreover, there are on average twice as many seeds Aa than AA. The self-fertilization of plants grown from the former gives seeds with either the dominant character (AA or Aa) or the recessive character (aa). As for the self-fertilization of plants grown from seeds AA, it always gives seeds AA with the dominant character. All the observations are thus explained.

Mendel also looked at the following generations. Starting from N hybrid seeds Aa and assuming for simplicity that each plant gives by self-fertilization only four new seeds, he computed that the mean number of seeds $(AA)_n$, $(Aa)_n$ and $(aa)_n$ in generation n would be given by Table 8.2, where for clarity of presentation the results have been divided by N.

Table 8.2 Successive generations.

n	0	1	2	3	4	5
$(AA)_n$	0	1	6	28	120	496
$(Aa)_n$	1	2	4	8	16	32
$(aa)_n$	0	1	6	28	120	496
total	1	4	16	64	256	1024

These numbers are simply obtained from the formulas

$$(AA)_{n+1} = (Aa)_n + 4(AA)_n , \tag{8.1}$$

$$(Aa)_{n+1} = 2(Aa)_n , \tag{8.2}$$

$$(aa)_{n+1} = (Aa)_n + 4(aa)_n , \tag{8.3}$$

which say that AA gives after self-fertilization four seeds AA, that aa gives four seeds aa and that Aa gives on average one seed AA, two seeds Aa and one seed aa. Mendel noticed furthermore that

$$(AA)_n = (aa)_n = 2^{n-1}(2^n - 1) \quad \text{and} \quad (Aa)_n = 2^n.$$

Indeed, it follows from equation (8.2) and from the initial condition $(Aa)_0 = 1$ that $(Aa)_n = 2^n$. Replacing this in equation (8.1), we get that $(AA)_{n+1} = 4(AA)_n + 2^n$. We easily realize that $(AA)_n = c\,2^n$ is a particular solution when $c = -1/2$. The general solution of the "homogeneous" equation $(AA)_{n+1} = 4(AA)_n$ is $(AA)_n = C\,4^n$. Finally, adding these two solutions, we see that $(AA)_n = C\,4^n - 2^{n-1}$ satisfies the initial condition $(AA)_0 = 0$ if $C = 1/2$. As for the sequence $(aa)_n$, it satisfies the same recurrence relation and the same initial condition as $(AA)_n$. So $(aa)_n = (AA)_n$. Q.E.D.

In conclusion, the proportion of hybrids Aa in the total population, which is $2^n/4^n = 1/2^n$, is divided by two at each generation by self-fertilization.

Mendel's work went totally unnoticed during his life. Some years later, Mendel also tried similar experiments with other plant species, published a few articles on meteorology and investigated the heredity of bees. After becoming abbot in 1868, he spent most of his time managing administrative problems. He died in 1884.

It is only in 1900 that Mendel's work was finally rediscovered independently and almost simultaneously by Hugo De Vries in Amsterdam, Carl Correns in Tübingen and Erich von Tschermak in Vienna. This would start a new era in what we now call genetics.

Further reading

1. Bateson, W.: *Mendel's Principles of Heredity*. Cambridge University Press (1913). www.archive.org
2. Mendel, J.G.: Versuche über Pflanzenhybriden. *Verh. Naturforsch. Ver. Brünn* **4**, 3–47 (1866). www.esp.org. English translation in Bateson (1913)
3. Fisher, R.A.: Has Mendel's work been rediscovered? *Ann. Sci.* **1**, 115–137 (1936). digital.library.adelaide.edu.au

Chapter 9
Galton, Watson and the extinction problem (1873–1875)

Francis Galton was born in 1822, the same year as Mendel, near Birmingham in England. He was the youngest of seven children. His father was a rich banker. Through his mother, he was the cousin of Charles Darwin. Galton started to study medicine in 1838, first in a hospital in Birmingham and later in London. During the summer of 1840, he made his first long trip through Europe as far as Istanbul. He then studied at Trinity College, Cambridge University, for four years. But his father died in 1844, leaving a significant fortune. Galton gave up the idea of becoming a medical doctor. He traveled to Egypt, Sudan and Syria. During the next few years he kept a leisured way of life, spending his time hunting, traveling in balloons and boats or trying to improve the electric telegraph. In 1850 he set up an expedition of exploration to South West Africa (now Namibia). On his return to England in 1852, he was elected to the Royal Geographical Society. There he could follow the news from the expeditions to Eastern Africa looking for the source of the Nile. He settled in London and wrote a guide book for travelers which became a best seller. In 1856 he was elected to the Royal Society. He was then interested in meteorology and invented the word "anticyclone". After the publication in 1859 by his cousin Darwin of *The Origin of Species*, Galton turned to the study of heredity. He published *Hereditary Genius* in 1869, in which he claimed that intellectual faculties could be transmitted by heredity.

In 1873 Alphonse de Candolle, a Swiss botanist, published a book entitled *History of Science and of Scientists in the Last Two Centuries*, which contained also an essay on "The Respective Influence of Heredity, Variability and Selection on the Development of the Human Species and on the Probable Future of this Species". There he made the following remarks:

> Among the precise piece of information and very sane opinions of the Mr Benoiston de Châteauneuf, Galton and other statisticians, I did not see the important remark they should have made on the *unavoidable* extinction of family names. Of course, every name has to go extinct [...] A mathematician could compute how the decrease of the names or titles would happen, knowing the probability of having female or male children and the probability of having no child for any given couple.

N. Bacaër, *A Short History of Mathematical Population Dynamics*,
DOI 10.1007/978-0-85729-115-8_9, © Springer-Verlag London Limited 2011

Fig. 9.1   Galton (left) and Watson (right).

It is the same problem that Bienaymé had studied in 1845. But Candolle, who was not aware of Bienaymé's work, thought that all families were bound to extinction. Galton noticed the above paragraph in Candolle's book. As he also did not know about Bienaymé's work, Galton put it as an open problem for the readers of the *Educational Times*:

> Problem 4001: A large nation, of whom we will only concern ourselves with adult males, N in number, and who each bear separate surnames colonise a district. Their law of population is such that, in each generation, a_0 per cent of the adult males have no male children who reach adult life; a_1 have one such male child; a_2 have two; and so on up to a_5 who have five.
>
> Find (1) what proportion of their surnames will have become extinct after r generations; and (2) how many instances there will be of the surname being held by m persons.

Notice that the second part of the problem had not been addressed by Bienaymé. Galton did not receive any satisfactory answer from the readers of the journal and apparently could not find the solution of the problem by himself. So he asked his friend Henry William Watson, a mathematician, to try to solve it.

Watson was born in London in 1827. His father was an officer in the British Navy. He first studied at King's College in London and then turned to mathematics at Trinity College, Cambridge University, from 1846 till 1850, just a few years after Galton. He became successively fellow of Trinity College, assistant master at the City of London School, lecturer in mathematics at King's College and professor of mathematics at Harrow School between 1857 and 1865. Fond of alpinism, he was part of an expedition which reached the top of Mount Rosa in Switzerland in 1855. He was ordained as a deacon in 1856 and as an Anglican priest two years later. From 1865 until his retirement he was rector of Berkswell with Barton near Coventry, a position which left enough time for research.

Galton and Watson wrote together an article entitled "On the probability of extinction of families", which was published in 1875 in the *Journal of the Royal*

Anthropological Institute. Galton presented the problem and Watson explained his computations and the conclusions he had reached. They assumed that men have at most q sons, p_k being the probability of having k sons ($k = 0, 1, 2, \ldots, q$). In other words, $p_k = a_k/100$ if we use Galton's original notations. So $p_0 + p_1 + \cdots + p_q = 1$. Consider the situation where generation 0 consists of a single man. Generation 1 consists of s men with a probability p_s. Using a trick which was well known in his time and which had been introduced long before by Abraham de Moivre, Watson considered the generating function

$$f(x) = p_0 + p_1 x + p_2 x^2 + \cdots + p_q x^q \qquad (9.1)$$

associated with the probabilities $p_0, \ldots, p_q$. Similarly, let $f_n(x)$ be the polynomial for which the coefficient of x^s is the probability of having s males in generation n starting from one man in generation 0. Then $f_1(x) = f(x)$. Watson noticed that

$$f_n(x) = f_{n-1}(f(x)), \qquad (9.2)$$

a formula which allows to compute $f_n(x)$ recursively.

Indeed, set $f_n(x) = p_{0,n} + p_{1,n} x + p_{2,n} x^2 + \cdots + p_{q^n,n} x^{(q^n)}$. Notice that there are at most q^n men in generation n. If in generation $n-1$ there are s men numbered 1 to s, call $t_1, \ldots, t_s$ the number of their male offspring. In such a case, there will be t men in generation n with a probability equal to

$$\sum_{t_1 + \cdots + t_s = t} p_{t_1} \times \cdots \times p_{t_s} .$$

When $s = 0$, it should be understood that this probability is equal to 1 if $t = 0$ and equal to 0 if $t \geq 1$. Therefore

$$p_{t,n} = \sum_{s \geq 0} p_{s,n-1} \times \sum_{t_1 + \cdots + t_s = t} p_{t_1} \times \cdots \times p_{t_s} .$$

It follows that

$$f_n(x) = \sum_{t \geq 0} p_{t,n} x^t = \sum_{s \geq 0} p_{s,n-1} \sum_{t \geq 0} \sum_{t_1 + \cdots + t_s = t} (p_{t_1} x^{t_1}) \times \cdots \times (p_{t_s} x^{t_s})$$

$$= \sum_{s \geq 0} p_{s,n-1} \left[p_0 x^0 + p_1 x^1 + p_2 x^2 + \cdots \right]^s$$

$$= \sum_{s \geq 0} p_{s,n-1} [f(x)]^s = f_{n-1}(f(x)) .$$

In particular the probability x_n of extinction of the family name within n generations is equal to $p_{0,n}$, which is the same as $f_n(0)$. As a first example, Watson took

$$f(x) = (1+x+x^2)/3,$$

i.e. $q = 3$ and $p_0 = p_1 = p_2 = 1/3$. He computed the polynomials $f_n(x)$ for $n = 1,\ldots,4$ using Eq. (9.2). He obtained for instance

$$f_2(x) = \frac{1}{3}\left[1 + \frac{1+x+x^2}{3} + \left(\frac{1+x+x^2}{3}\right)^2\right] = \frac{13 + 5x + 6x^2 + 2x^3 + x^4}{27}$$

and $f_2(0) = 13/27 \simeq 0.481$. The computation of $f_n(x)$ for $n \geq 3$ becomes very tedious, so tedious that Watson already made a mistake for $n = 4$. Since $x_5 = f_5(0) = f_4(f(0))$ can avoid the computation of $f_5(x)$, he got the following list of extinction probabilities $x_n = f_n(0)$:

$$x_1 \simeq 0.333, \quad x_2 \simeq 0.481, \quad x_3 \simeq 0.571, \quad x_4 \simeq 0.641, \quad x_5 \simeq 0.675.$$

The correct values are $x_4 \simeq 0.632$ and $x_5 \simeq 0.677$, as can be checked by using the simple formula $x_n = f(x_{n-1})$ derived by Bienaymé. As we shall see in Chapter 17, the latter formula can also be derived from Eq. (9.2).

Watson noticed that each man has on average $\mathcal{R}_0 = p_1 + 2p_2 + \cdots + qp_q$ sons and that $\mathcal{R}_0 = 1$ in his first example. So one could think that if the initial number of male family members was large enough, the family size would remain roughly constant. Nevertheless Watson claimed that the extinction probability x_n converges towards 1 when $n \to +\infty$, though quite slowly. In other words the whole family will reach extinction as Candolle had suggested. Figure 9.2a, which is not drawn in the original article, and Bienaymé's results confirm that this conclusion for the first example is correct.

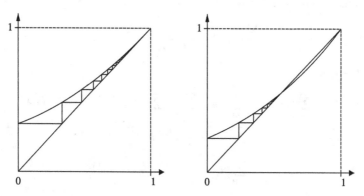

Fig. 9.2 Graph of the functions $y = f(x)$ and $y = x$. The extinction probability $x_n = f(x_{n-1})$ within n generations is the height of the nth "step of the staircase". Left: $f(x) = (1+x+x^2)/3$. Right: $f(x) = (3+x)^5/4^5$.

As a second example, Watson considered the binomial probability distribution

$$p_k = \binom{q}{k} \frac{a^{q-k}b^k}{(a+b)^q}, \tag{9.3}$$

for which the generating function (9.1) is $f(x) = (a+bx)^q/(a+b)^q$. He computed $f_2(x)$ and $x_2 = f_2(0)$. At this point he realized that $x_2 = f(x_1)$ and that $x_n = f(x_{n-1})$ for all n. But he thought this formula was true only for the special binomial case (9.3). Applying it to the case where $q = 5$, $a = 3$ and $b = 1$, he obtained

$$x_1 \simeq 0.237, \quad x_2 \simeq 0.347, \quad x_3 \simeq 0.410, \quad \ldots \quad x_9 \simeq 0.527, \quad x_{10} \simeq 0.533, \ldots$$

Watson realized that x_n has to converge to a limit x_∞ as $n \to +\infty$, which satisfies $x_\infty = f(x_\infty) = (a+bx_\infty)^q/(a+b)$. He noticed that $x = 1$ is a solution of this equation but did not realize that there could be other solutions when $\mathscr{R}_0 > 1$. So he erroneously concluded, misled by Candolle, that there is extinction ($x_\infty = 1$) in every case, including the numerical example he had just considered. Fig. 9.2b shows that this it not case!

Watson noticed that the mean number of sons in this numerical example was bigger than 1 (one can show that $\mathscr{R}_0 = qb/(a+b) = 5/4$), meaning that the population tends to increase exponentially. But this did not help him discover his error. He even conjectured that extinction of the family name was certain for every probability distribution (p_k), i.e. not just for the binomial case. We shall return to this problem in Chapters 17 and 18.

Galton continued his statistical study of families with a book entitled *English Men of Science, their Nature and Nurture*, which focused on the genealogy of fellows of the Royal Society. He also became interested in anthropometry, the measure of the human body. He took advantage of an international exhibit in 1884 in London to collect data on a large number of people. His results were published in 1889 in a book entitled *Natural Inheritance*, whose appendix reproduced the article written in collaboration with Watson. This book also introduced some new statistical vocabulary such as "percentile" and "quartile" as well as the word "eugenics", i.e. the improvement of the human species from the point of view of hereditary characters. After 1888 Galton developed the technique of recognizing fingerprints that was to be used a few years later by the British police. He also continued to study the respective role of heredity (nature) and of the environment (nurture) on physical and intellectual characteristics of twins, on the size of peas grown over several generations or on the color of mice bred in a laboratory. This led him to the notion of "correlation coefficient" between two variables. In 1904 the Galton Laboratory was founded within University College in London. Galton was knighted in 1909 and died in 1911.

Watson published several books, in particular a treatise on the kinetic theory of gases in 1876 and a treatise on the mathematical theory of electricity and magnetism in two volumes (1885 and 1889). He was elected to the Royal Society in 1881 and died in Brighton in 1903.

In 1924, in the second volume of his biography of Galton, Karl Pearson summarized the article on the extinction of family names without noticing the error. This error would finally be noticed in 1930 (see Chapter 18).

Further reading

1. De Candolle, A.: *Histoire des sciences et des savants depuis deux siècles suivie d'autres études sur des sujets scientifiques en particulier sur la sélection dans l'espèce humaine*. Georg, Genève (1873). www.archive.org
2. Galton, F.: *Natural Inheritance*. Macmillan, London (1889). galton.org
3. Galton, F.: *Memories of my Life*. Methuen & Co., London (1908). galton.org
4. Kendall, D.G.: Branching processes since 1873. *J. Lond. Math. Soc.* **41**, 385–406 (1966)
5. Pearson, K.: *The Life, Letters and Labours of Francis Galton*, vol. 1. Cambridge University Press (1914). galton.org
6. Pearson, K.: *The Life, Letters and Labours of Francis Galton*, vol. 2. Cambridge University Press (1924). galton.org
7. S.H.B.: Henry William Watson, 1827-1903. *Proc. R. Soc. Lond.* **75**, 266–269 (1905). gallica.bnf.fr
8. Watson, H.W., Galton, F.: On the probability of the extinction of families. *J. Anthropol. Inst.* **4**, 138–144 (1875). galton.org

Chapter 10
Lotka and stable population theory (1907–1911)

Alfred James Lotka was born of American parents in 1880 in Lemberg, which was part of the Austro-Hungarian Empire (now L'viv in Ukraine). He studied first in France and Germany and in 1901 obtained a bachelor's degree in physics and chemistry from the University of Birmingham in England. He then spent one year in Leipzig where the role of thermodynamics in chemistry and biology was emphasized by Wilhelm Ostwald, who was to receive the Nobel prize in Chemistry in 1909. Lotka settled in New York in 1902 and began to work for the General Chemical Company.

Fig. 10.1 Lotka (1880–1949)

In 1907 and 1911[1], Lotka took up the study of the dynamics of age-structured populations without knowing about Euler's work on the same subject (see Chapter 3). Unlike Euler he assumed that time and age are continuous variables. Let $B(t)$ be the male birth rate (the number of male births per unit of time) at time t, $p(x)$ the probability of being still alive at age x and $h(x)$ the fertility at age x: $h(x)\,dx$ is

---

[1] The second article was written in collaboration with F.R. Sharpe, a mathematician from Cornell University.

N. Bacaër, *A Short History of Mathematical Population Dynamics*,
DOI 10.1007/978-0-85729-115-8_10, © Springer-Verlag London Limited 2011

the probability for a man to have one newborn son between age x and $x + dx$ if dx is infinitely small. Then

$$\int_0^\infty p(x)\,dx$$

is the life expectancy at birth. Moreover $B(t-x)\,p(x)\,dx$ is the number of males born between time $t-x$ and $t-x+dx$, which are still alive at time t. These males have $B(t-x)\,p(x)\,h(x)\,dx$ sons per unit of time at time t. So the total male birth rate at time t is

$$B(t) = \int_0^\infty B(t-x)\,p(x)\,h(x)\,dx.$$

Looking for an exponential solution for this integral equation in the unknown $B(t)$ of the form $B(t) = b\,e^{rt}$, Lotka obtained by dividing both sides by $B(t)$ the equation

$$1 = \int_0^\infty e^{-rx}\,p(x)\,h(x)\,dx, \tag{10.1}$$

which is now called "Lotka's equation" by demographers[2]. Euler had obtained the analogous implicit equation (3.1) for the growth rate when time and age are discrete variables. Lotka noticed that the right-hand side of (10.1) is a decreasing function of r which tends to $+\infty$ when $r \to -\infty$ and which tends to 0 when $r \to +\infty$. So there is a unique value of r, call it r^*, such that equation (10.1) holds. Besides, $r^* > 0$ if and only if

$$\mathscr{R}_0 = \int_0^\infty p(x)\,h(x)\,dx > 1. \tag{10.2}$$

The parameter $\mathscr{R}_0$ (the notation was introduced by Dublin and Lotka in 1925) is the expected number of sons that one man may have throughout his life.

Lotka suggested[3] that, whatever the initial age structure of the population, the number of male births per unit of time was indeed such that $B(t) \sim b\,e^{r^*t}$ when $t \to +\infty$, where b is a constant. The total population is then given by

$$P(t) = \int_0^\infty B(t-x)\,p(x)\,dx.$$

It follows that $P(t)$ also increases or decreases like e^{r^*t} when $t \to +\infty$: the growth rate is equal to r^*. Moreover, the population's age structure, given by $B(t-x)\,p(x)/P(t)$, tends to

$$\frac{e^{-r^*x}\,p(x)}{\int_0^\infty e^{-r^*y}\,p(y)\,dy}.$$

This is what Lotka called a "stable population": the age pyramid keeps the same shape through time but the total population increases or decreases exponentially.

---

[2] R.A. Fisher arrived independently at the same equation in 1927 and later interpreted the root r^* as a measure of "Darwinian fitness" in the theory of evolution by natural selection.

[3] This was rigorously proven in 1941 by Willy Feller, who was then professor of mathematics at Brown University in the USA. A probabilistic approach was developed in 1968 by Crump, Mode and Jagers.

The conclusion is thus the same as in Euler's discrete-time model. But Lotka's study takes into account the age dependence of fertility. So it is in some sense more general than Euler's.

Lotka continued to work on this topic throughout his life. In 1908–1909 he resumed his studies at Cornell University to get a master's degree. He worked for the National Bureau of Standards from 1909 till 1911 and as editor of the journal *Scientific American Supplement* from 1911 till 1914. In 1912 he obtained a doctorate from the University of Birmingham by collecting the articles he had published since 1907 on population dynamics and demography. During the First World War, he worked again for the General Chemical Company on how to fix nitrogen from the atmosphere. In 1920 one of his articles on biological oscillations (see Chapter 13) made a deep impression on Raymond Pearl, a professor of biometry at Johns Hopkins University who had just "rediscovered" the logistic equation (see Chapter 6). Hoping to find a job at the Rockefeller Institute of Medical Research in New York, Lotka worked on the mathematical models developed by Ross for malaria (see Chapter 12). Finally he got a two-year scholarship from Johns Hopkins University, which allowed him to write a book entitled *Elements of Physical Biology*, published in 1925. He then became the head of the research department of the Metropolitan Life Insurance Company in New York. He focused on the mathematical analysis of demographic questions and published several books in collaboration with a colleague, the statistician and vice-president of the company Louis Israel Dublin: *The Money Value of a Man* (1930), *Length of Life* (1936) and *Twenty Five years of Health Progress* (1937). He was elected president of the Population Association of America for 1938–1939. Among his various statistical studies, "Lotka's law" (going back to 1926) states that the number of authors having written n articles in a given scientific field decreases more or less like $1/n^2$ as n increases.

Lotka also published a book in French entitled *Analytical Theory of Biological Associations*. The first part, which was more philosophical, appeared in 1934. The second more technical part, published in 1939, summarized all his research on human demography since 1907. In his book Lotka also presented his contribution to the problem of extinction of family names. After the publication in 1930 of Steffensen's first article on the subject (see Chapter 18), he had applied the theory to the data contained in the 1920 census of the white population of the USA. He noticed that the observed distribution $(p_k)_{k \geq 0}$ of the number of sons is well approximated by a decreasing geometric law for all $k \geq 1$:

$$p_0 = a, \quad p_k = b c^{k-1} \ (k \geq 1),$$

with $a = 0.4825$, $b = 0.2126$ and $c = 1 - b/(1 - a)$. In this way, $\sum_{k \geq 0} p_k = 1$. The associated generating function is

$$f(x) = a + b \sum_{k=1}^{+\infty} c^{k-1} x^k = a + \frac{bx}{1 - cx}.$$

The two solutions of the equation $x = f(x)$ are $x = 1$ and $x = a/c$. The extinction probability x_∞ is the smallest of these two solutions (see Chapter 7). With the numerical values for the USA he found $x_\infty \simeq 0.819$, while the mean number of sons was $\mathscr{R}_0 = f'(1) = (1 - a)^2/b \simeq 1.260$. Despite a mean number of children (including sons and daughters) close to 2.5, the probability of extinction of the family name is above 80%.

Lotka was elected president of the American Statistical Association in 1942. He retired in 1947 and died in 1949 in New Jersey. A new edition of his 1925 book appeared in 1956 with the slightly different title *Elements of Mathematical Biology*.

Further reading

1. Crump, K.S., Mode, C.J.: A general age-dependent branching process. *J. Math. Anal. Appl.* **24**, 494–508 (1968)
2. Dublin, L.I., Lotka, A.J.: On the true rate of natural increase. *J. Amer. Stat. Assoc.* **20**, 305–339 (1925)
3. Feller, W.: On the integral equation of renewal theory. *Ann. Math. Stat.* **12**, 243–267 (1941). projecteuclid.org
4. Fisher, R.A.: The actuarial treatment of official birth records. *Eugen. Rev.* **19**, 103–108 (1927). digital.library.adelaide.edu.au
5. Gridgeman, N.T.: Lotka, Afred James. In: Gillespie, C.C. (ed.) *Dictionary of Scientific Biography*, vol. 8, p. 512. Scribner, New York (1981)
6. Kingsland, S.E.: *Modeling Nature, Episodes in the History of Population Ecology*, 2nd edn. University of Chicago Press (1995). books.google.com
7. Lotka, A.J.: Relation between birth rates and death rates. *Science* **26**, 21–22 (1907) Reprinted in Smith & Keyfitz (1977)
8. Lotka, A.J.: *Théorie analytique des associations biologiques*, 2ième partie. Hermann, Paris (1939) English translation: *Analytical Theory of Biological Populations*. Plenum Press, New York (1998). books.google.com
9. Sharpe, F.R., Lotka, A.J.: A problem in age-distribution. *Philos. Mag. Ser.* 6, **21**, 435–438 (1911) Reprinted in Smith & Keyfitz (1977)
10. Smith, D.P., Keyfitz, N.: *Mathematical Demography, Selected Papers*. Springer, Berlin (1977)

Chapter 11
The Hardy–Weinberg law (1908)

Godfrey Harold Hardy was born in 1877 in Surrey, England. His parents were teachers. He studied mathematics at Trinity College, Cambridge University, from 1896, became a fellow of his college in 1900 and a lecturer in mathematics in 1906. After a first book on *The Integration of Functions of a Single Variable* (1905), he published in 1908 *A Course of Pure Mathematics*, which was reedited many times and translated to many foreign languages.

Fig. 11.1 Hardy (1877–1947)

At that time, the rediscovery of Mendel's work had raised some doubts. Some biologists wondered why the dominant characters did not become more frequent from generation to generation. Reginald Punnett, who had written a book entitled *Mendelism* in 1905, asked the question to Hardy, with whom he played cricket in Cambridge. Hardy wrote his solution in an article on "Mendelian proportions in a mixed population", which was published in 1908. To simplify the analysis, he imagined the situation of a large population where the choice of the sexual partner would be random. Moreover he restricted his attention to just two factors (or "alleles") A and a, A being dominant and a recessive. For generation n, let p_n be the frequency

N. Bacaër, *A Short History of Mathematical Population Dynamics*,
DOI 10.1007/978-0-85729-115-8_11, © Springer-Verlag London Limited 2011

of the "genotype" AA, $2q_n$ that of Aa and r_n that of aa. Of course, $p_n + 2q_n + r_n = 1$. Hardy assumed also that none of these genotypes led to an excess of mortality or to a decrease in fertility when compared to the two other genotypes. The frequencies in generation $n+1$ can be easily computed by noticing that one randomly chosen individual in generation n transmits allele A with a probability $p_n + q_n$: either the genotype is AA and allele A is transmitted for sure or the genotype is Aa and allele A is transmitted with 50% chance. Similarly, allele a is transmitted with a probability $q_n + r_n$. One can hence construct Table 11.1 in the same way as Table 8.1.

Table 11.1 Computation of the frequencies of the genotypes in generation $n+1$ from the frequencies of the alleles of the parents (rows are for the mother, columns for the father).

Allele	A	a
Frequency	$p_n + q_n$	$q_n + r_n$
A	AA	Aa
$p_n + q_n$	$(p_n + q_n)^2$	$(p_n + q_n)(q_n + r_n)$
a	Aa	aa
$q_n + r_n$	$(p_n + q_n)(q_n + r_n)$	$(q_n + r_n)^2$

The frequencies of the genotypes AA, Aa and aa in generation $n+1$ are respectively p_{n+1}, $2q_{n+1}$ and r_{n+1}. So Hardy found that

$$p_{n+1} = (p_n + q_n)^2 \tag{11.1}$$
$$2q_{n+1} = 2(p_n + q_n)(q_n + r_n) \tag{11.2}$$
$$r_{n+1} = (q_n + r_n)^2. \tag{11.3}$$

He then investigated under which conditions the frequencies of the genotypes could stay constant through the generations being equal to p, $2q$ and r. Since by definition $p + 2q + r = 1$, we see that equations (11.1)-(11.3) all yield the same condition $q^2 = pr$.

For example, the first equation gives $p = (p+q)^2 = p^2 + 2pq + q^2$, which is equivalent to $p(1 - p - 2q) = q^2$ and finally to $pr = q^2$.

Starting from arbitrary initial conditions $(p_0, 2q_0, r_0)$ with $p_0 + 2q_0 + r_0 = 1$, Hardy noticed that
$$q_1^2 = (p_0 + q_0)^2 (q_0 + r_0)^2 = p_1 r_1.$$

The state $(p_1, 2q_1, r_1)$ is therefore already an equilibrium. So $(p_n, 2q_n, r_n)$ stays equal to $(p_1, 2q_1, r_1)$ for all $n \geq 1$. If we set $x = p_0 + q_0$ for the frequency of allele A in generation 0, then $1 - x = q_0 + r_0$ is the frequency of allele a. Using system (11.1)–(11.3) once again, we get

$$p_n = x^2, \quad 2q_n = 2x(1-x), \quad r_n = (1-x)^2$$

for all $n \geq 1$ (Fig. 11.2).

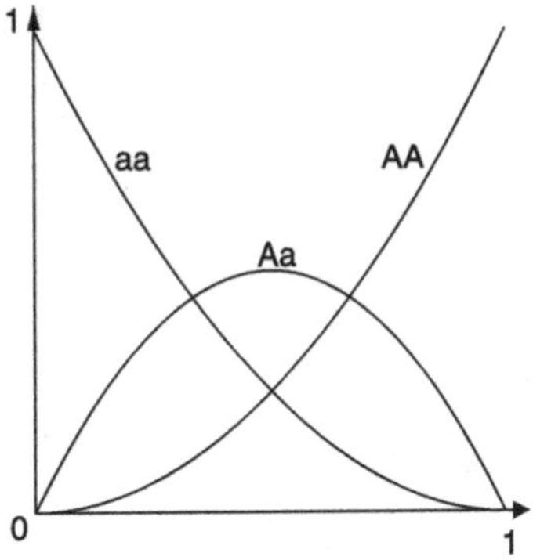

Fig. 11.2 Graphs of the functions x^2, $2x(1-x)$ and $(1-x)^2$ corresponding to the equilibrium frequencies of the genotypes AA, Aa and aa.

In conclusion, the above hypotheses lead to the law according to which *the frequencies of the genotypes AA, Aa and aa stay unchanged through generations.* Mendel's theory does not lead to a progressive increase of the frequency of the dominant character as had first been thought.

Some years later, Fisher would insist on an important corollary of this law: to a first approximation (i.e. assuming that the model's hypotheses are realistic), a population keeps a constant genetic variance. This observation solves one of the problems raised by Darwin's theory of evolution by natural selection. Indeed, Darwin thought, like his contemporaries, that at each generation the physiological characteristics of the children were a kind of average of the characteristics of the two parents, each parent contributing one half. This idea had been later thoroughly studied using statistics by Francis Galton and his successor at the biometry laboratory, Karl Pearson. If it were true, the variance of these characteristics in a population should be divided by two at each generation and there would soon be such a homogeneity that natural selection, supposed to explain evolution, would be impossible. Several years would nevertheless be necessary for this averaging mechanism to be rejected, biometricians defending Darwin's point of view and being reluctant to admit that Mendel's laws are unavoidable to understand evolution.

After this work in 1908, Hardy returned to pure mathematics. In his autobiography, *A Mathematician's Apology*, he even claimed with pride having avoided discoveries of any practical use. In 1910 he was elected to the Royal Society. In 1913 he discovered the Indian prodigy Ramanujan and invited him to work in Cambridge. After the First World War, he became professor at Oxford University and continued a fruitful collaboration with his compatriot Littlewood. Between 1931 and 1942 he was again professor in Cambridge. He published many books, often in collabora-

tion: *Orders of Infinity* (1910), *The General Theory of Dirichlet's Series* with Marcel
Riesz (1915), *Inequalities* with Littlewood and Pólya (1934), *An Introduction to the
Theory of Numbers* with E. M. Wright (1938), *Ramanujan* (1940), *Fourier Series*
with Rogosinski (1944) and *Divergent Series* (1949). He died in Cambridge in 1947.

Fig. 11.3 Weinberg (1862–1937)

Several decades later, people noticed that Hardy's law for gene frequencies had
also been discovered that same year 1908 by a German medical doctor, Wilhelm
Weinberg. Weinberg was born in Stuttgart in 1862. After studying in Tübingen and
Munich until his doctorate in medicine, he had worked several years in hospitals in
Berlin, Vienna and Frankfurt. He had settled in 1889 in Stuttgart as general practi-
tioner and obstetrician. Despite being very busy with his work, he had found time
to write many articles in German scientific journals. In 1901 he had studied from a
statistical point of view the frequency of twins of the same sex. The 1908 article,
in which he explained the same law as Hardy had found, had been published in a
local scientific journal and had not been noticed. But unlike Hardy, he had continued this study the following years, discovering for example the generalization to the
case where there are more than two alleles. He had also contributed to the area of
medical statistics. Weinberg died in 1937. After the rediscovery of his 1908 article,
geneticists called the law of stability of genotype frequencies the "Hardy–Weinberg
law".

Nowadays this law is often used as follows. If a rare recessive allele a has no
influence on survival or fertility and if we know the frequency x^2 of the genotype aa
because aa produces a particular phenotype, then we can compute x and estimate
the frequency $2x(1-x) \simeq 2x$ of the genotype Aa. As an example, if the frequency
of aa is $1/20,000$, then we get $x \simeq 1/140$. So $2x \simeq 1/70$ is the frequency of the
genotype Aa. The recessive allele a, which might appear very rare from inspection
of the phenotypes, is in fact not so uncommon.

Further reading

1. Hardy, G.H.: Mendelian proportions in a mixed population. *Science* **28**, 49–50 (1908). www.esp.org
2. Hardy, G.H.: *A Mathematician's Apology*. Foreword by C.P. Snow. Cambridge University Press (1967). books.google.com
3. Punnett, R.C.: *Mendelism*, 2nd edn. Cambridge University Press (1907). www.archive.org
4. Stern, C.: The Hardy–Weinberg law. *Science* **97**, 137–138 (1943)
5. Stern, C.: Wilhelm Weinberg 1862–1937. *Genetics* **47**, 1–5 (1962)
6. Titchmarsh, E.C.: Godfrey Harold Hardy, 1877–1947. *Obit. Not. Fellows R. Soc.* **6**, 446–461 (1949)
7. Weinberg, W.: Über den Nachweis der Vererbung beim Menschen. *Jahresh. Wuertt. Ver. vaterl. Natkd.* **64**, 369–382 (1908). www.biodiversitylibrary.org

Chapter 12
Ross and malaria (1911)

Ronald Ross was born in 1857 in the North of India, where his father was an officer in the British army. He studied medicine in London but preferred to write poems and dramas. After working for a year on a ship as surgeon, he managed to enter the Indian Medical Service in 1881. His medical work in India left him plenty of free time, during which he wrote literary works and taught himself some mathematics. On furlough to England in 1888, he obtained a diploma in public health and studied bacteriology, a new science created a few years earlier by Pasteur and Koch. Back in India, Ross started to study malaria. During his second furlough in 1894, he met in London Patrick Manson, a specialist in tropical medicine who showed him under the microscope what the French military doctor Alphonse Laveran had noticed in 1880: the blood of patients with malaria contains parasites. Manson suggested that the parasites could come from mosquitoes because he had discovered himself in China the parasite of another tropical disease (filariasis) in these insects. However, he believed that humans were infected by the parasite when drinking water contaminated by the mosquitoes. From 1895 till 1898, Ross continued his research in India and tested Manson's idea. In 1897 he discovered in the stomach of a certain mosquito species that he had not studied before (anopheles) some parasites similar to those observed by Laveran. His superiors having sent him to Calcutta during a season where malaria cases were rare, he decided to study malaria in cage birds. He found the parasite in the salivary glands of anopheles mosquitoes and managed to infect experimentally healthy birds by letting mosquitoes bite them: this proved that malaria is transmitted by mosquito bites and not by ingestion of contaminated water. In 1899 Ross left the Indian Medical Service to teach at the Liverpool School of Tropical Medicine, which had been created one year before. He was elected to the Royal Society in 1901 and received in 1902 the Nobel Prize in Physiology or Medicine for his work on malaria. He traveled to Africa, to Mauritius and in the Mediterranean area to popularize the fight against mosquitoes. The method was successful in Egypt along the Suez canal, along the Panama canal under construction, in Cuba and in Malaysia. It was less successful in some other areas. Ross published a *Report on the Prevention of Malaria in Mauritius* in 1908 and *The Prevention of Malaria* in 1910.

N. Bacaër, *A Short History of Mathematical Population Dynamics*,
DOI 10.1007/978-0-85729-115-8_12, © Springer-Verlag London Limited 2011

Fig. 12.1 Ross (1857–1932)

Despite his proof of the role of certain mosquitoes in the transmission of malaria, Ross met skepticism when he claimed that malaria could be eradicated simply by reducing the number of mosquitoes. In the second edition of his book on *The Prevention of Malaria* published in 1911, he tried to build mathematical models of the transmission of malaria in order to support his claim. One of his models consisted of a system of two differential equations. Let us introduce the following notations:

- N: total human population in a given area;
- $I(t)$: number of humans infected with malaria at time t;
- n: total mosquito population (assumed constant);
- $i(t)$: number of mosquitoes infected with malaria;
- b: biting frequency of mosquitoes;
- p (respectively p'): transmission probability of malaria from human to mosquito (respectively from mosquito to human) during one bite;
- a: rate at which humans recover from malaria;
- m: mosquito mortality.

During a small time interval dt, each infected mosquito bites $b\,dt$ humans, among which a fraction equal to $\frac{N-I}{N}$ is not yet infected. Taking into account the transmission probability p', there are $b\,p'\,i\,\frac{N-I}{N}\,dt$ new infected humans. During the same time interval, the number of humans that recover is $a I\,dt$. Hence,

$$\frac{dI}{dt} = b\,p'\,i\,\frac{N-I}{N} - aI.$$

Similarly each noninfected mosquito bites $b\,dt$ humans, among which a fraction equal to I/N is already infected. Taking into account the transmission probability p, there are $b\,p\,(n-i)\,\frac{I}{N}\,dt$ new infected mosquitoes. Meanwhile, assuming that infection does not influence mortality, the number of mosquitoes that die is $m\,i\,dt$. So

$$\frac{di}{dt} = b\,p\,(n-i)\,\frac{I}{N} - mi.$$

Since malaria exists permanently in most infected countries, Ross considered just the steady states of his system of two equations: the number of infected humans $I(t)$ and the number of infected mosquitoes $i(t)$ stay constant through time ($dI/dt = 0$ and $di/dt = 0$). First there is always the steady state with $I = 0$ and $i = 0$, which corresponds to the absence of malaria. Secondly, Ross looked for a steady state such that $I > 0$ and $i > 0$ and found that

$$I = N\frac{1 - amN/(b^2 p p' n)}{1 + aN/(b p' n)}, \quad i = n\frac{1 - amN/(b^2 p p' n)}{1 + m/(b p)}. \qquad (12.1)$$

Dividing the steady state equations by the product $I \times i$, the problem becomes a linear system of two equations with two unknowns $1/I$ and $1/i$,

$$\frac{b p'}{I} - \frac{a}{i} = \frac{b p'}{N}, \quad -\frac{m}{I} + \frac{b p n}{N i} = \frac{b p}{N}.$$

Its solution is easily obtained.

One can notice that $I > 0$ and $i > 0$ if the number of mosquitoes is above a critical threshold:

$$n > n^* = \frac{amN}{b^2 p p'}.$$

In this case the steady state corresponds to the situation where the disease is endemic, i.e. permanently present. Ross concluded that if the number of mosquitoes n is reduced below the critical threshold n^*, then the only remaining steady state is $I = 0$ and $i = 0$, so malaria should disappear. In particular it is not necessary to exterminate *all* the mosquitoes to eradicate malaria. This is precisely the point Ross wanted to emphasize with his model.

To illustrate his theory, Ross looked for reasonable numerical values for the parameters of his model. He assumed that

- the mortality of mosquitoes is such that only one third of them are still alive after ten days; so $e^{-10m} = \frac{1}{3}$ and $m = (\log 3)/10$ per day;
- half of the people are still infected after three months; so $e^{-90a} = 1/2$ and $a = (\log 2)/90$ per day;
- one out of eight mosquitoes bites each day; so $e^{-b} = 1 - 1/8$ and $b = \log(8/7)$ per day;
- infected mosquitoes are usually not infectious during the first ten days following their infection because the parasites have to go through several stages of transformation. Since one third of mosquitoes can survive ten days, Ross assumed that there are also about one third of all infected mosquitoes that are infectious: $p' = 1/3$;
- $p = 1/4$.

Ross could then compute with formula (12.1) the infected fraction I/N in the human population as a function of the ratio n/N between the mosquito and the human population. He showed his results in a table that is equivalent to Fig. 12.2.

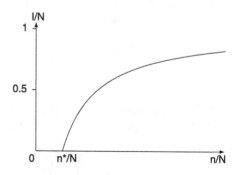

Fig. 12.2 Fraction I/N of infected humans as a function of the ratio n/N between the mosquito and the human population.

The shape of the curve shows that the fraction of infected humans is higher than 50% already if the ratio n/N is just slightly above the critical value n^*/N. But this fraction does not change much when the ratio n/N increases further. This explains why the correlation between the number of mosquitoes and the presence of malaria had never been noticed before. Ross noticed, however, that the numerical value of the threshold n^*/N was very sensitive to small changes in the biting rate b, but that this did not change the overall shape of the curve in Fig. 12.2. His qualitative explanation is more important than the quantitative results, which suffer from the uncertainty surrounding the numerical values of the parameters.

To interpret the critical threshold n^* discovered by Ross[1], consider one infected human introduced in a human and a mosquito population both free of malaria. This human stays infected on average during a period of time equal to $1/a$. He or she receives bn/N bites per unit of time so on average $bn/(aN)$ bites in total while infected. So he or she infects on average $bpn/(aN)$ mosquitoes. Each of these infected mosquitoes lives on average during a period of time equal to $1/m$, bites b/m humans and infects bp'/m humans. In total, after the transmission from the first infected human to the mosquitoes and from these mosquitoes to other humans, the mean number of newly infected humans is the product of the previous two results, i.e.

$$\mathscr{R}_0 = \frac{b^2 p p' n}{amN}.\qquad(12.2)$$

This $\mathscr{R}_0$ is the number of secondary human cases due to one primary human case. So the infection process which happens continuously in time can also be considered through successive generations. Malaria can "invade" the population only if $\mathscr{R}_0 > 1$. This condition is precisely equivalent to $n > n^*$.

---

[1] This interpretation was emphasized only long after Ross' work.

In conclusion, Ross pleaded more generally in favour of mathematical modelling in epidemiology:

> As a matter of fact all epidemiology, concerned as it is with the variation of disease from time to time or from place to place, *must* be considered mathematically, however many variables are implicated, if it is to be considered scientifically at all. To say that a disease depends upon certain factors is not to say much, until we can also form an estimate as to how largely each factor influences the whole result. And the mathematical method of treatment is really nothing but the application of careful reasoning to the problems at issue.

Ross was knighted in 1911. He moved to London and became a consultant for the British army during the First World War. In 1923 he published his autobiography, *Memoirs With a Full Account of the Great Malaria Problem and its Solution*. In 1926 was inaugurated the Ross Institute of Tropical Diseases (now part of the London School of Hygiene and Tropical Medicine), of which he became the director. Ross died in London in 1932.

Further reading

1. G.H.F.N.: Sir Ronald Ross, 1857–1932. *Obit. Not. Fellows Roy. Soc.* **1**, 108–115 (1933)
2. Ross, R.: *The Prevention of Malaria*, 1st edn. John Murray, London (1910). www.archive.org
3. Ross, R.: *The Prevention of Malaria*, 2nd edn. John Murray, London (1911)
4. Ross, R.: *Memoirs with a Full Account of the Great Malaria Problem and its Solution*. John Murray, London (1923). www.archive.org
5. Rowland, J.: *The Mosquito Man, The Story of Sir Ronald Ross*. Roy Publishers, New York (1958)

Chapter 13
Lotka, Volterra and the predator–prey system (1920–1926)

In 1920 Lotka published an article entitled "Analytical note on certain rhythmic relations in organic systems". For some years already, he had been interested in some
chemical reactions that exhibited strange transitory oscillations in laboratory experiments. The purpose of his article was to suggest that a system of two biological
species could even oscillate permanently. The example he considered was that of
a population of herbivores feeding on plants. In analogy with the equations used
in chemical kinetics, let $x(t)$ be the total mass of plants and $y(t)$ the total mass of
herbivores at time t. Lotka used as a model the following system of differential
equations

$$\frac{dx}{dt} = ax - bxy, \tag{13.1}$$

$$\frac{dy}{dt} = -cy + dxy, \tag{13.2}$$

where the parameters a, b, c and d are all positive. Parameter a is the growth rate of
plants when there are no herbivores, while c is the rate of decrease of the population
of herbivores when there are no plants. The terms $-bxy$ and dxy express that the
more animals and plants there are, the higher the mass transfer from plants towards
animals (the transfer includes some loss of mass so $d \leq b$). Setting $dx/dt = 0$ and
$dy/dt = 0$, Lotka noticed that there are two steady states:

- $(x = 0, y = 0)$, the population of herbivores is extinct and there are no more
  plants;
- $(x = c/d, y = a/b)$, herbivores and plants coexist.

He also wrote without proof that if at time $t = 0$, $(x(0), y(0))$ is not one of these
two steady states, then the functions $x(t)$ and $y(t)$ oscillate periodically: there is a
number $T > 0$ such that $x(t+T) = x(t)$ and $y(t+T) = y(t)$ for all $t > 0$ (Fig. 13.1)[1].
If for example the plants are very abundant, then the population of herbivores will
increase, causing a decrease in the total mass of plants. When this mass becomes

---

[1] The period T depends on the initial conditions, but Lotka realized this fact only in 1925.

N. Bacaër, *A Short History of Mathematical Population Dynamics*,
DOI 10.1007/978-0-85729-115-8_13, © Springer-Verlag London Limited 2011

insufficient to feed the herbivores, some animals die of hunger and the total mass of plants will start to grow again until it reaches a level equal to its initial value. The phenomenon will repeat itself.

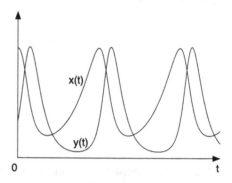

Fig. 13.1 Oscillations of the total mass of plants $x(t)$ and of the total mass of herbivores $y(t)$ as a function of time.

Lotka studied the model a little further in a second article published in 1920 entitled "Undamped oscillations derived from the law of mass action". He explained why the system could oscillate in a periodic way. This follows from the fact that the point $(x(t),y(t))$ has to stay on a closed trajectory in the plane with x on the horizontal axis and y on the vertical axis; more precisely, in the quadrant where $x \geq 0$ and $y \geq 0$ (Fig. 13.2).

Indeed, dividing equation (13.1) by equation (13.2), we obtain after some re-ordering

$$\left(-\frac{c}{x}+d\right)\frac{dx}{dt} = \left(\frac{a}{y}-b\right)\frac{dy}{dt}.$$

Integration gives

$$dx(t) - c\log x(t) = by(t) - a\log y(t) + K,$$

where K is a constant which depends only on the initial condition. Hence, the point $(x(t),y(t))$ stays on the curve $dx - c\log x = by - a\log y + K$, which happens to be a closed curve (Fig. 13.2).

The trajectory of $(x(t),y(t))$ turns around the steady state $(c/d,a/b)$ counter-clockwise as can be easily seen by studying the sign of dx/dt and of dy/dt. Near the steady state, the system exhibits small oscillations with a period equal to $2\pi/\sqrt{ac}$.

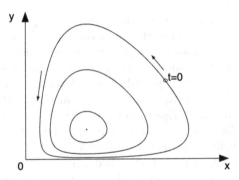

Fig. 13.2 Diagram with the total mass of plants $x(t)$ on the horizontal axis and the total mass of herbivores $y(t)$ on the vertical axis. The three closed curves around the steady state correspond to different initial conditions.

Indeed, set $x = \frac{c}{d} + x^*$ and $y = \frac{a}{b} + y^*$ where $x^* \ll \frac{c}{d}$ and $y^* \ll \frac{a}{b}$. Then

$$\frac{dx^*}{dt} = -by^*\left(\frac{c}{d} + x^*\right) \simeq -\frac{bc}{d}y^*,$$

$$\frac{dy^*}{dt} = dx^*\left(\frac{a}{b} + y^*\right) \simeq \frac{ad}{b}x^*.$$

From these two equations, we obtain

$$\frac{d^2x^*}{dt^2} \simeq -acx^* \quad \text{and} \quad \frac{d^2y^*}{dt^2} \simeq -acy^*.$$

These equations are the same as for the oscillations of the simple pendulum in physics. The period is $2\pi/\sqrt{ac}$.

Raymond Pearl, who had communicated the first 1920 article to the *Proceedings of the National Academy of Sciences*, helped Lotka get a two-year scholarship from Johns Hopkins University to write a book entitled *Elements of Physical Biology*. The book was published in 1925. The section summarizing the 1920 work also mentioned that systems of two species, one host and one parasite species or one prey and one predator species, could be described by the same model (13.1)–(13.2). Unfortunately Lotka's book did not draw much attention when it was published. However, the famous mathematician Volterra independently rediscovered that same model soon after while studying a fishery problem.

Vito Volterra was born in the Jewish ghetto of Ancona in 1860, shortly before the unification of Italy, when the city still belonged to the Papal States. He was a single child. His father, a cloth merchant, died when Vito was two years old and left the family without money. A good student in high school, Volterra managed to continue studying despite poverty, first at the University of Florence and later at the Scuola Normale Superiore in Pisa. In 1882 he obtained a doctorate in physics and the following year became professor of mechanics at the University of Pisa. He joined the University of Turin in 1892 and moved to a chair of mathematical physics at the University La Sapienza in Rome in 1900. He became senator in 1905. Many of the lectures he gave in Rome or in foreign universities were published in book form: *Three Lessons on Some Recent Progress in Mathematical Physics* (Clark University, 1909), *Lessons on Integral and Integro-differential Equations* (Rome, 1910), *Lessons on Line Functions* (Paris, 1912), *The Theory of Permutable Functions* (Princeton, 1912). He served as an officer in the Italian army during the First World War and led the bureau of war inventions. After the war, he participated actively in the foundation of the Italian Mathematical Union (1922) and of the Italian National Research Council (1923), becoming the first chairman of the latter. He also became president of the International Commission for the Scientific Study of the Mediterranean Sea (1923) and president of the Accademia dei Lincei (1924). Another monograph, written in collaboration with J. Pérès, *Lessons on Composition and Permutable Functions*, was published in 1924.

Fig. 13.3 Volterra (1860–1940) receiving a doctorate *honoris causa* from the University of Cambridge in 1900.

In 1925, at age 65, Volterra became interested in a study by the zoologist Umberto D'Ancona, who would later become his son-in-law, on the proportion of cartilaginous fish (such as sharks and rays) landed in the fishery during the years 1905–1923 in three harbours of the Adriatic Sea: Trieste, Fiume[2] and Venice. D'Ancona had noticed that the proportion of these fish had increased during the First World War, when the fishing effort had been reduced (Table 13.1).

The cartilaginous fish being predators of smaller fish, it seemed that a decrease in the fishing effort favoured the predator species. Volterra, who did not know about Lotka's work, explained this observation by using the same model

---

[2] Now Rijeka in Croatia.

Table 13.1 Percentage of cartilaginous fish in the fisheries of Trieste, Fiume and Venice before, during and after the First World War.

Year	1910	1911	1912	1913	1914	1915	1916	1917	1918	1919	1920	1921	1922	1923
Trieste	5.7	8.8	9.5	15.7	14.6	7.6	16.2	15.4	-	19.9	15.8	13.3	10.7	10.2
Fiume	-	-	-	-	11.9	21.4	22.1	21.2	36.4	27.3	16.0	15.9	14.8	10.7
Venice	21.8	-	-	-	-	-	-	-	-	30.9	25.3	25.9	25.8	26.6

$$\frac{dx}{dt} = ax - bxy, \quad \frac{dy}{dt} = -cy + dxy,$$

where $x(t)$ stands for the number of prey and $y(t)$ for the number of predators. He noticed, like Lotka, that this system can oscillate in a periodic way with a period T which depends on the initial condition (x_0, y_0). He noticed also that

$$\frac{d}{dt} \log x = a - by, \quad \frac{d}{dt} \log y = -c + dx.$$

Integrating over one period T (so that $x(0) = x(T)$ and $y(0) = y(T)$), he obtained

$$\frac{1}{T} \int_0^T y(t)\, dt = \frac{a}{b}, \quad \frac{1}{T} \int_0^T x(t)\, dt = \frac{c}{d}.$$

So the average over one period of both the number of prey and the number of predators is independent of the initial conditions. Moreover, if the fishing effort decreases, the growth rate a of prey increases while the extinction rate c of predators decreases. Therefore the average of $x(t)$ decreases and the average of $y(t)$ increases: the proportion of predators increases. This is precisely what had been observed for the fishery statistics from the Adriatic Sea.

Volterra published his article first in Italian in 1926. An English summary appeared a few months later in *Nature*. Lotka informed Volterra and other scientists of the priority of his study of predator–prey systems. But his 1920 article and his 1925 book would not always be mentioned. Lotka was then already working for an insurance company, so his work focused on human demography. Volterra continued to work on variants of the predator–prey system for a decade. He gave a series of lectures in 1928-1929 at the newly created Institut Henri Poincaré in Paris. The notes of these lectures were published in 1931 under the title *Lessons on the Mathematical Theory of the Struggle for Life*. In 1935 Volterra published in collaboration with Umberto D'Ancona another book on *Biological Associations from a Mathematical Point of View*.

Although the predator–prey model seems to explain the fishery data correctly, the debate concerning the realism of simplified models in ecology was just starting and is still a subject of scientific dispute. Nowadays, the predator–prey model is also known as the Lotka– Volterra model and is one the most commonly cited in ecology.

In 1931 Volterra refused to give allegiance to Mussolini. He lost his professorship at the university in Rome and was excluded from Italian scientific academies, of which he was one of the most famous members. From then on he remained mainly outside Italy, travelling through Europe and giving lectures. He published with J. Pérès the first volume of a *General Theory of Functionals* (1936) and a book with B. Hostinský on *Infinitesimal Linear Operations* (1938). He died in Rome in 1940.

Further reading

1. Goodstein, J.R.: *The Volterra Chronicles, The Life and Times of an Extraordinary Mathematician 1860–1940*. American Mathematical Society (2007). books.google.com

2. Guerraggio, A., Nastasi, P.: *Italian Mathematics between the Two World Wars*. Birkhäuser, Basel (2005). books.google.com

3. Israel, G., Gasca, A.M.: *The Biology of Numbers – The Correspondence of Vito Volterra on Mathematical Biology*. Birkhäuser, Basel (2002)

4. Kingsland, S.E.: *Modeling Nature, Episodes in the History of Population Ecology*, 2nd edn. University of Chicago Press (1995). books.google.com

5. Lotka, A.J.: Analytical note on certain rhythmic relations in organic systems. *Proc. Natl. Acad. Sci.* **6**, 410–415 (1920). www.pnas.org

6. Lotka, A.J.: Undamped oscillations derived from the law of mass action. *J. Amer. Chem. Soc.* **42**, 1595–1599 (1920). www.archive.org

7. Lotka, A.J.: *Elements of Physical Biology*. Williams & Wilkins, Baltimore (1925). www.archive.org

8. Volterra, V.: Variazioni e fluttuazioni del numero d'individui in specie animali conviventi. *Mem. Accad. Lincei* **6**, 31–113 (1926) Reprinted in: *Opere matematiche*, vol. 5, Accademia nazionale dei Lincei, Roma (1962)

9. Volterra, V.: Fluctuations in the abundance of a species considered mathematically. *Nature* **118**, 558–560 (1926). Reprinted in L.A. Real, J.H. Brown (eds.) *Foundations of Ecology*, pp. 283–285. University of Chicago Press (1991)

10. Volterra, V.: *Leçons sur la Théorie Mathématique de la Lutte pour la Vie*. Gauthier-Villars, Paris (1931)

11. Volterra, V., D'Ancona, U.: *Les Associations Biologiques au Point de Vue Mathématique*. Hermann, Paris (1935)

12. Whittaker, E.T.: Vito Volterra 1860–1940. *Obit. Not. Fellows R. Soc.* **3**, 690–729 (1941)

Chapter 14
Fisher and natural selection (1922)

Ronald Aylmer Fisher was born in London in 1890, the last of six children. His father was an auctioneer, but later declared bankruptcy. Fisher studied mathematics and physics at Gonville and Caius College of Cambridge University between 1909 and 1913. Genetics was developing quickly at the time. Starting in 1911, Fisher participated in the meetings of the Eugenics Society initiated by Galton. He started to focus on statistical problems related to the work of Galton and Mendel. After finishing his university studies he spent one summer working on a farm in Canada and then worked for the Mercantile and General Investment Company in the City of London. Because of his extreme shortsightedness, he could not participate in the First World War despite having volunteered. He spent these years teaching in high schools. During his free time, he took care of a farm and continued his research. He obtained important new results linking correlation coefficients with Mendelian genetics. In 1919 he started to work as a statistician at the Rothamsted Experimental Station, which focused on agriculture.

Fig. 14.1 Fisher (1890–1962)

N. Bacaër, *A Short History of Mathematical Population Dynamics*,
DOI 10.1007/978-0-85729-115-8_14, © Springer-Verlag London Limited 2011

In 1922 Fisher published an article entitled "On the dominance ratio". Among several other important new ideas, this article considered a mathematical model combining Mendel's laws and the idea of natural selection emphasized by Darwin for the theory of evolution. Fisher considered the same situation as Hardy with two alleles A and a and with the random mating hypothesis. But he assumed that individuals with genotypes AA, Aa and aa have different mortalities before reaching adulthood, thus mimicking natural selection. Setting p_n, $2q_n$ and r_n for the frequencies of the three genotypes among adult individuals in generation n, there are respectively $(p_n + q_n)^2$, $2(p_n + q_n)(q_n + r_n)$ and $(q_n + r_n)^2$ newborns in generation $n + 1$ having these genotypes. Let u, v and w be the respective survival probabilities from birth to adulthood. Then the frequencies of the genotypes among adult individuals in generation $n + 1$ are p_{n+1}, $2q_{n+1}$ and r_{n+1} with

$$p_{n+1} = \frac{u(p_n + q_n)^2}{d_n} \tag{14.1}$$

$$q_{n+1} = \frac{v(p_n + q_n)(q_n + r_n)}{d_n} \tag{14.2}$$

$$r_{n+1} = \frac{w(q_n + r_n)^2}{d_n}, \tag{14.3}$$

where we set for convenience

$$d_n = u(p_n + q_n)^2 + 2v(p_n + q_n)(q_n + r_n) + w(q_n + r_n)^2.$$

Remembering that $p_n + 2q_n + r_n = 1$, we see that when $u = v = w$ (i.e. when there is no natural selection), system (14.1)-(14.3) reduces to the system (11.1)-(11.3) considered by Hardy.

Let $x_n = p_n + q_n$ be the frequency of allele A among adult individuals in generation n. Then $q_n + r_n = 1 - x_n$ is the frequency of allele a. Adding (14.1) and (14.2), we get

$$x_{n+1} = \frac{u x_n^2 + v x_n(1 - x_n)}{u x_n^2 + 2v x_n(1 - x_n) + w(1 - x_n)^2}.$$

This equation can be rewritten in the form

$$x_{n+1} - x_n = x_n(1 - x_n) \frac{(v - w)(1 - x_n) + (u - v)x_n}{u x_n^2 + 2v x_n(1 - x_n) + w(1 - x_n)^2}. \tag{14.4}$$

There are always at least two steady states where the frequency x_n stays constant through generations: $x = 0$ (the population consists entirely of homozygous aa) and $x = 1$ (the population consists entirely of homozygous AA).

Using equation (14.4), one can show that if the homozygous AA has a better chance of survival than the two other genotypes ($u > v$ and $u > w$), then allele a will progressively disappear from the population. This case should not be very common in nature if we know that both alleles coexist. If, however, the heterozygous Aa has a selective advantage over the homozygous AA and aa ($v > u$ and $v > w$), then the three genotypes can coexist in the population. This is the most common case and it can explain the "vigour" of hybrids noticed by farmers.

Indeed, the steady state $x = 1$ is stable when $u > v$ because $x_{n+1} - x_n \simeq (1 - x_n)(u - v)/u$ when x_n is close to 1. The population tends to this steady state. The steady state $x = 1$ is unstable when $u < v$, in which case there is a third steady state

$$x^* = \frac{v - w}{2v - u - w}$$

with $0 < x^* < 1$. Moreover we can check that this one is stable. The steady state x^* corresponds to a mixture between the three genotypes.

Hence combining simply Mendel's laws and a hypothesis of natural selection (here, different survival probabilities for the three genotypes), we can explain the two situations of coexistence or disappearance of genotypes. After Fisher, this model was also developed by J.B.S. Haldane (see Chapter 17) and by Sewall Wright (see Chapter 19).

In anticipation of Chapter 20, notice that if A is completely dominant and the homozygous aa is disadvantaged compared to the two other genotypes, the numbers $u : v : w$ being in a ratio $1 : 1 : 1 - \varepsilon$, then equation (14.4) becomes

$$x_{n+1} - x_n = \frac{\varepsilon x_n (1 - x_n)^2}{1 - \varepsilon(1 - x_n)^2} \simeq \varepsilon x_n (1 - x_n)^2 \tag{14.5}$$

for $\varepsilon \ll 1$. If the survival of the heterozygous Aa lies halfway between that of the two homozygous, then the numbers $u : v : w$ are in a ratio $1 : 1 - \varepsilon/2 : 1 - \varepsilon$ and

$$x_{n+1} - x_n = \frac{\frac{\varepsilon}{2} x_n (1 - x_n)}{1 - \varepsilon(1 - x_n)} \simeq \frac{\varepsilon}{2} x_n (1 - x_n) \tag{14.6}$$

when $\varepsilon \ll 1$.

At Rothamsted Fisher analysed long-term data concerning crop yields and meteorology. But he also made great contributions to statistical methodology. In 1925 he published a book entitled *Statistical Methods for Research Workers*, which was highly successful and reprinted many times. He became a fellow of the Royal Society in 1929. In 1930 Fisher published a book on *The Genetical Theory of Natural Selection*, a milestone in the history of population genetics. He became professor of eugenics at University College in London in 1933, succeeding Karl Pearson at the Galton Laboratory. In 1943 he moved to a genetics chair at Cambridge University, this time succeeding R.C. Punnett (see Chapter 11). He also published several books: *The Design of Experiments* (1935), *The Theory of Inbreeding* (1949) and *Statistical Methods and Scientific Inference* (1956). Knighted in 1952, he settled in

Australia after retiring in 1959 and died in Adelaide in 1962. We shall return to another part of his work in Chapter 20.

Further reading

1. Fisher Box, J.: R.A. Fisher, *The Life of a Scientist*. John Wiley & Sons, New York (1978)
2. Fisher, R.A.: On the dominance ratio. *Proc. R. Soc. Edinb.* **42**, 321–341 (1922). digital.library.adelaide.edu.au
3. Fisher, R.A.: *The Genetical Theory of Natural Selection*. Clarendon Press, Oxford (1930). www.archive.org
4. Yates, F., Mather, K.: Ronald Aylmer Fisher 1890–1962. *Biog. Mem. Fellows R. Soc.* **9**, 91–120 (1963). digital.library.adelaide.edu.au

Chapter 15
Yule and evolution (1924)

George Udny Yule was born in Scotland in 1871, his father having had a high level position in the British administration in India. At the age of 16 Yule started to study at University College in London to become an engineer. In 1892 he changed his orientation and spent one year doing research in Bonn under the supervision of the physicist Heinrich Hertz, who had demonstrated the existence of electromagnetic waves a few years earlier. When Yule returned to England, Karl Pearson offered him a position of assistant professor in applied mathematics at University College. Yule, following Pearson, began to focus on statistics. In 1911 he published *An Introduction to the Theory of Statistics*, which was reprinted 14 times. The following year he moved to Cambridge University. His research work dealt with theoretical aspects of statistics but also with applications to agriculture and epidemiology. He became a fellow of the Royal Society in 1922.

Fig. 15.1 Yule (1871–1951)

In 1924 Yule published an article entitled *A mathematical theory of evolution based on the conclusions of Dr. J. C. Willis*. Willis was a colleague from the Royal

N. Bacaër, *A Short History of Mathematical Population Dynamics*,
DOI 10.1007/978-0-85729-115-8_15, © Springer-Verlag London Limited 2011

Society who had published in 1922 a book entitled *Age and Area, A Study in Geographical Distribution and Origin of Species*. He had studied the distribution of species among different genera in the classification of plants and animals. The data that he had compiled showed that most genera contained only one species, that fewer and fewer genera contained higher numbers of species and that there were still a few genera containing a great number of species. Table 15.1 shows the data concerning snakes, lizards and two families of beetles (the Chrysomelidae and the Cerambycinae). The 1,580 species of lizards known at the time had been classified in 259 genera, 105 genera containing only one species, 44 only two species, 23 only three species, etc., and two genera containing more than one hundred species. For other families of animals and plants, the distribution of genera according to the number of species they contain had a very similar shape. Yule suggested that Willis should try to plot his data in a graph with logarithmic scales. This gave a striking result (Fig. 15.2): the logarithm of the number Q_n of genera containing n species decreases more or less linearly with $\log(n)$. In other words, there are constants $\alpha > 0$ and $\beta > 0$ such that $Q_n \simeq \alpha n^{-\beta}$: the distribution follows a "power law". In his 1924 article, Yule looked for a mathematical model of evolution that could explain such a statistical distribution.

Table 15.1 Data compiled by Willis.

| Number of species | Number of genera | | | |
	Chrysomelidae	Cerambycinae	Snakes	Lizards
1	215	469	131	105
2	90	152	35	44
3	38	82	28	23
4	35	61	17	14
5	21	33	16	12
6	16	36	9	7
7	15	18	8	6
8	14	17	8	4
9	5	14	9	5
10	15	11	4	5
11-20	58	74	10	17
21-30	32	21	12	9
31-40	13	15	3	3
41-50	14	8	1	2
51-60	5	4	0	0
61-70	8	3	0	1
71-80	7	0	1	0
81-90	7	1	0	0
91-100	3	1	1	0
101-	16	4	0	2
total	627	1024	293	259

Fig. 15.2 The number of
genera as a function of
the number of species they
contain, with decimal logarithmic scales. Data for the
Chrysomelidae. To smooth
the fluctuations when n
(the number of species) is
large, genera were counted
for ranges of n-values as in
Tab. 15.1. The average number of genera for a single
value of n can thus be less
than 1.

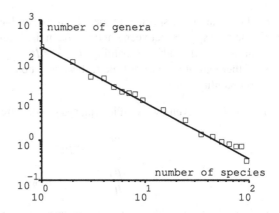

For this purpose he imagined first a continuous-time stochastic model[1] for the
growth of the number of species within one genus (Fig. 15.3). Starting with only
one species at time $t = 0$, he assumed that the probability for a species to give birth
by mutation to a new species of the same genus during a "small" time interval dt
(on the time scale of evolution) was equal to $r\,dt$ with $r > 0$.

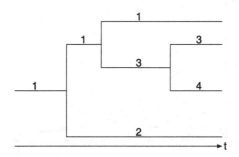

Fig. 15.3 A simulation of the evolution of the number of species within one genus. Species 1
generates species 2 and 3. Species 3 generates species 4.

Let $p_n(t)$ be the probability that there are n species at time t (n is an integer but t
is a real number). To compute $p_n(t + dt)$, Yule considered several cases:

---

[1] McKendrick (see Chapter 16) had already started to study such models in population dynamics
in a paper published in 1914.

- if there are $n-1$ species at time t, each species has a probability $r\,dt$ of generating one new species between t and $t+dt$; in the limit $dt \to 0$, there will be n species at time $t+dt$ with a probability $(n-1)r\,dt$;
- if there are n species at time t, there will be $n+1$ species at time $t+dt$ with a probability $nr\,dt$.

Thus $p_n(t)$ is given by the following system of differential equations

$$\frac{dp_1}{dt} = -r p_1, \tag{15.1}$$

$$\frac{dp_n}{dt} = (n-1)r p_{n-1} - nr p_n \tag{15.2}$$

for all $n \geq 2$. From the first equation, we get $p_1(t) = e^{-rt}$ because $p_1(0) = 1$. It is possible to show that the solution of the second equation that satisfies the initial condition $p_n(0) = 0$ is

$$p_n(t) = e^{-rt}(1 - e^{-rt})^{n-1} \tag{15.3}$$

for all $n \geq 2$ (Fig. 15.4). So at some fixed time t, the distribution of probabilities $(p_n(t))_{n\geq1}$ is "geometric" with a ratio between two consecutive terms equal to $1 - e^{-rt}$.

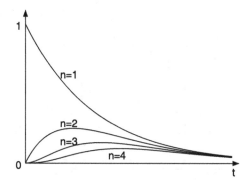

Fig. 15.4 The probability $p_n(t)$ that there are n species of the same genus at time t, for $1 \leq n \leq 4$.

Indeed, we notice first that equation (15.2) is equivalent to

$$\frac{d}{dt}\left[p_n e^{nrt} \right] = (n-1)r p_{n-1} e^{nrt}, \tag{15.4}$$

from which we can compute successively $p_2(t)$, $p_3(t)$ .... We get $p_2(t) = e^{-rt}(1 - e^{-rt})$, then $p_3(t) = e^{-rt}(1 - e^{-rt})^2$, which suggests formula (15.3) for the general solution. One can finally check that this formula is a solution of equation (15.4).

Yule also deduced from formula (15.3) that the expected number of species increases exponentially with time:

$$\sum_{n=1}^{\infty} n\, p_n(t) = e^{rt}.$$

Indeed, we notice first that for $|x| < 1$,

$$\sum_{n=1}^{\infty} n x^{n-1} = \frac{d}{dx} \sum_{n=0}^{\infty} x^n = \frac{d}{dx}\left(\frac{1}{1-x}\right) = \frac{1}{(1-x)^2}.$$

Then

$$\sum_{n=1}^{\infty} n\, p_n(t) = e^{-rt} \sum_{n=1}^{\infty} n(1 - e^{-rt})^{n-1} = e^{rt}.$$

In particular, if T is the doubling time defined by $e^{rT} = 2$, then the probability distribution $(p_n(t))_{n \geq 1}$ of the number of species at time $t = T$ is geometric with a ratio $1/2$:

$$\frac{1}{2}, \quad \frac{1}{4}, \quad \frac{1}{8}, \quad \frac{1}{16} \quad \cdots$$

At time $t = kT$, it is geometric with a ratio $1 - 1/2^k$ and $p_1(kT) = 1/2^k$.

Yule next considered, in parallel to the growth of the number of species belonging to the same genus, a similar process due to larger mutations leading to the creation of new genera. Let $s\,dt$ be the probability for an existing genus to generate a new genus during a small time interval dt. As before, assuming that there is only one genus at time $t = 0$, the expected number of genera at time t is e^{st}. The mean number of genera created per unit of time at time t is the derivative $s e^{st}$. In the limit[2] where $t \to +\infty$, the mean number of genera which at time t have existed between x and $x + dx$ units of time is then $s e^{s(t-x)}\,dx$. The probability at time t for a randomly chosen genus to have existed between x and $x + dx$ units of time is $s e^{-sx}\,dx$.

If a genus chosen at random at time t has existed between x and $x + dx$ units of time, the probability that this genus contains n species is, according to formula (15.3), equal to $e^{-rx}(1 - e^{-rx})^{n-1}$ for all $n \geq 1$. So the probability q_n for a genus randomly chosen at time t to contain n species is

---

[2] Yule considered also the case where t cannot be assumed very large compared to the doubling time of e^{st}. The computations are a little more complicated but the final results are not very different.

$$q_n = \int_0^\infty s e^{-sx} e^{-rx} (1 - e^{-rx})^{n-1} dx.$$

Set $u = r/s$. An easy computation shows that $q_1 = 1/(1+u)$ and that

$$q_n = \frac{1}{1+u} \frac{u}{1+2u} \frac{2u}{1+3u} \cdots \frac{(n-1)u}{1+nu} \qquad (15.5)$$

for all $n \geq 2$.

Indeed, we have $(1 - e^{-rx})^{n-1} = (1 - e^{-rx})^{n-2} (1 - e^{-rx})$. So

$$q_n = q_{n-1} - s \int_0^\infty e^{-(r+s)x} (1 - e^{-rx})^{n-2} e^{-rx} dx.$$

Integrating by parts, we get

$$q_n = q_{n-1} - \frac{r+s}{(n-1)r} q_n \quad \text{and} \quad q_n = \frac{(n-1)r/s}{1+nr/s} q_{n-1}.$$

Formula (15.5) shows that the sequence of probabilities $(q_n)_{n \geq 1}$ is decreasing. So the maximum is reached for $n = 1$: most genera contain just one species. This is precisely what the data had shown. Moreover, the decrease of q_n towards 0 when n tends to infinity is relatively slow because $q_n/q_{n-1} \to 1$. This may explain why some genera contain a large number of species. More precisely, Yule showed that $\log q_n$ decreases linearly with $\log(n)$.

Introduce Euler's Gamma function $\Gamma(z) = \int_0^\infty t^{z-1} e^{-t} dt$. Then $\Gamma(n+1) = n! = n \times (n-1) \times \cdots \times 2 \times 1$ when n is integer and $\Gamma(z+1) = z\Gamma(z)$. So (15.5) takes the form

$$q_n = \frac{(n-1)!}{u(1+\frac{1}{u})(2+\frac{1}{u})\cdots(n+\frac{1}{u})} = \frac{\Gamma(n)\Gamma(1+\frac{1}{u})}{u\,\Gamma(n+1+\frac{1}{u})}.$$

But Stirling's approximation gives $\log \Gamma(n) \simeq n\log n - n - \frac{1}{2}\log n + \text{constant}$. Similarly, $\log \Gamma(n+1+1/u) \simeq n\log n - n + (\frac{1}{u}+\frac{1}{2})\log n + \text{constant}$. Finally $\log q_n \simeq -(1+\frac{1}{u})\log n + \text{constant}$.

Consider for example the case of lizards. Parameter u can be estimated from the proportion $q_1 = 1/(1+u)$ of genera that contain only one species. According to Table 15.1, we have $q_1 = 105/259$ so $u \simeq 1.467$. We can then compute the theoretical probability q_n and the expected number Q_n of genera containing n species by multiplying q_n with the total number of species, which is 259 (Table 15.2). Yule noticed that the agreement between the observations and the computations is relatively good[3] given the model's simplicity, which does not take into account for example the cataclysms that species have crossed through millions of years of evolution.

Table 15.2 Comparison between data and theory in the case of lizards (1,580 species classified in 259 genera).

Number of species per genus	Observed number of genera	Computed number of genera
1	105	105.0
2	44	39.2
3	23	21.3
4	14	13.6
5	12	9.6
6	7	7.2
7	6	5.6
8	4	4.5
9	5	3.7
10	5	3.1
11-20	17	16.6
21-30	9	6.9
31-40	3	3.9
41-50	2	2.6
51-60	0	1.9
61-70	1	1.4
71-80	0	1.1
81-90	0	0.9
91-100	0	0.7
101-	2	10.1
total	259	259

After 1931 Yule retired progressively from Cambridge University. He became interested in the statistical distribution of the length of sentences to identify book authors. He applied this in particular to the book published by John Graunt (see Chapter 2) but possibly inspired by William Petty. In 1944 he published a book on *The Statistical Study of Literary Vocabulary*. He died in 1951.

Nowadays Yule's model is still used to analyze "phylogenetic trees" (the genealogical trees of species). These trees, similar to that in Fig. 15.3, are better known thanks to the new data coming from molecular biology. But the applications of the stochastic process defined by equations (15.1)-(15.2) are not limited to the theory of evolution. This process is a building block of many models in population dynamics, from the microscopic level (to model for example colonies of bacteria) to the macroscopic level (to model the beginning of an epidemic). It is called "pure

---

[3] For the number of genera containing more than 100 species, Yule got a better fit than in Table 15.2 by considering that t was not large compared to the doubling time of e^{st}.

birth process" or "Yule process". A simple variant includes a probability $m\,dt$ of dying during any small time interval dt: the expected population size at time t for this "birth and death process" is then $e^{(r-m)t}$. As for the probability distribution (15.5), it is sometimes called the Yule distribution . Distributions with tails satisfying power laws have attracted a lot of attention in various areas of science. The study of epidemics in random networks with a power law degree distribution is just one example.

Further reading

1. Aldous, D.J.: Stochastic models and descriptive statistics for phylogenetic trees, from Yule to today. *Stat. Sci.* **16**, 23–34 (2001). projecteuclid.org
2. Edwards, A.W.F.: George Udny Yule. In: Heyde, C.C., Seneta, E. (eds.) *Statisticians of the Centuries*, pp. 292–294. Springer, New York (2001)
3. McKendrick, A.G.: Studies on the theory of continuous probabilities with special reference to its bearing on natural phenomena of a progressive nature. *Proc. Lond. Math. Soc.* **13**, 401–416 (1914)
4. Simon, H.A.: On a class of skew distribution functions. *Biometrika* **42**, 425–440 (1955)
5. Willis, J.C.: *Age and Area, A Study in Geographical Distribution and Origin of Species.* Cambridge University Press (1922). www.archive.org
6. Yates, F.: George Udny Yule. *Obit. Not. Fellows R. Soc.* **8**, 308–323 (1952)
7. Yule, G.U.: A mathematical theory of evolution, based on the conclusions of Dr. J. C. Willis, F.R.S. *Phil. Trans. Roy. Soc. Lond. B* **213**, 21–87 (1925). gallica.bnf.fr

Chapter 16
McKendrick and Kermack on epidemic modelling (1926–1927)

Anderson Gray McKendrick was born in 1876 in Edinburgh, the last of five children. He studied medicine at the University of Glasgow where his father was a professor of physiology. In 1900 he joined the Indian Medical Service. Before going to India, he accompanied Ronald Ross on a mission to fight malaria in Sierra Leone. He then served in the army for 18 months in Sudan. At his arrival in India, he was appointed as medical doctor in a prison in Bengal where he tried to control dysentery. In 1905 he joined the new Central Institute for Medical Research in Kasauli (in the North of India). He worked on rabies but also studied mathematics. In 1920, having been infected by a tropical disease, he returned to Edinburgh and became the superintendent of the Royal College of Physicians Laboratory.

Fig. 16.1 McKendrick (1876–1943) and Kermack (1898-1970)

In 1926 McKendrick published an article on the "Applications of mathematics to medical problems", which contained several new ideas. He introduced in particular a continuous-time mathematical model for epidemics that took into account the stochastic aspect of infection and recovery.

N. Bacaër, *A Short History of Mathematical Population Dynamics*, DOI 10.1007/978-0-85729-115-8_16, © Springer-Verlag London Limited 2011

Consider a population of size N with initially only one infected person. People can go successively through three states: the susceptible state S, the infected state I and the recovered state R (Fig. 16.2)[1].

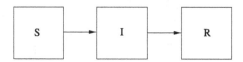

Fig. 16.2 Possible states: susceptible (S), infected (I), recovered (R).

Let $p_{i,r}(t)$ be the probability that the population contains at time t exactly i people in state I and r people in state R, where i and r are integers such that $1 \leq i+r \leq N$. In this case the population is said to be in state (i,r). The number of susceptible people is $s = N - i - r$. Following the work of Ross on malaria (see Chapter 12), McKendrick assumed that, during a small time interval dt, the probability for one new infection to occur is equal to $a s i\, dt$ (i.e. proportional to both the number of susceptible people and the number of infected people). The probability for one new recovery is equal to $b i\, dt$. Both a and b are positive parameters. To compute $p_{i,r}(t + dt)$, several cases should be distinguished:

- the population is in state $(i-1,r)$ at time t and one new infection moves the population to state (i,r) between t and $t+dt$; the probability of this event is $a s(i-1)\, dt$ with $s = N - (i-1) - r$;
- the population is in state (i,r) at time t and one new infection moves the population to state $(i+1,r)$ between t and $t+dt$; the probability of this event is $a s i\, dt$ with $s = N - i - r$;
- the population is in state $(i+1,r-1)$ at time t and one new recovery moves the population to state (i,r) between t and $t+dt$; the probability of this event is $b(i+1)\, dt$;
- the population is in state (i,r) at time t and one new recovery moves the population to state $(i-1,r+1)$ between t and $t+dt$; the probability of this event is $b i\, dt$.

Hence, McKendrick obtained the equations

$$\frac{dp_{i,r}}{dt} = a(N-i-r+1)(i-1)\, p_{i-1,r} - a(N-i-r)i\, p_{i,r}$$
$$+ b(i+1)\, p_{i+1,r-1} - b i\, p_{i,r} \tag{16.1}$$

for $1 \leq i+r \leq N$. The first term on the right-hand side is missing when $i = 0$, whereas the third term is missing when $r = 0$. The initial conditions are $p_{i,r}(0) = 0$ for all (i,r) except $p_{1,0}(0) = 1$.

---

[1] Daniel Bernoulli's model (see Chapter 4) included the states S and R but not I, the duration of infection being much shorter than the average life expectancy.

With this model McKendrick managed to compute the probability for the epidemic to end with n people having been infected, which is the limit of $p_{0,n}(t)$ when $t \to +\infty$. Indeed there is no need to solve system (16.1). It is enough to notice that as long as there are i infected people and r recovered people, the probability of a new infection during a small time interval dt is $a(N-i-r)i\,dt$ and the probability of a new recovery is $bi\,dt$. So the transition probabilities (as they are usually called in the theory of "Markov chains") from state (i,r) to state $(i+1,r)$ or state $(i-1,r+1)$ are respectively

$$\mathscr{P}_{(i,r)\to(i+1,r)} = \frac{a(N-i-r)}{a(N-i-r)+b}\,,\qquad \mathscr{P}_{(i,r)\to(i-1,r+1)} = \frac{b}{a(N-i-r)+b}\,,$$

for all $i \geq 1$ (Fig. 16.3).

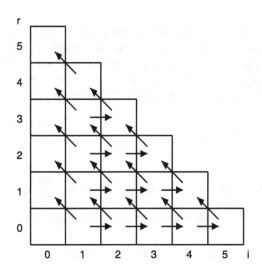

Fig. 16.3 Diagram showing the possible states of a population with $N = 5$ (i on the horizontal axis, r on the vertical axis) and the possible transitions due to infection (horizontal arrows) or to recovery (other arrows).

Let $q_{i,r}$ be the probability that the population goes through state (i,r) during the epidemic. Since $i = 1$ and $r = 0$ when $t = 0$, we have $q_{1,0} = 1$. The other states are reached either after an infection or after a recovery:

$$q_{i,r} = q_{i-1,r}\,\mathscr{P}_{(i-1,r)\to(i,r)} + q_{i+1,r-1}\,\mathscr{P}_{(i+1,r-1)\to(i,r)}\,.$$

The first term of the right-hand side is missing when $i = 0$ or $i = 1$. The second term is missing when $r = 0$. From this formula, we can first compute $(q_{i,0})_{2 \leq i \leq N}$, then $(q_{i,1})_{0 \leq i \leq N-1}$, then $(q_{i,2})_{0 \leq i \leq N-2}$ etc. The probability that the epidemic will finally infect n people is $q_{0,n}$. In 1926 such computations were quite tedious. So McKendrick limited himself to examples concerning very small populations, for example a family. With $N = 5$ people and $b/a = 2$, he obtained Table 16.1. The largest probabilities correspond to the case where only one person in the family is infected and to the case where the entire family is infected.

Table 16.1 Probability of an epidemic in a family of five to infect n people when $b/a = 2$.

n	1	2	3	4	5
$q_{0,n}$	0.33	0.11	0.09	0.13	0.34

The same article from 1926 contains also a new formulation of demographic problems when time is considered as a continuous variable. For dx infinitely small, let $P(x,t)\,dx$ be the population with an age between x and $x + dx$ at time t. Let $m(x)$ be the mortality at age x. Then $P(x+h, t+h) \simeq P(x,t) - m(x)P(x,t)h$ for h infinitely small. Introduce the partial derivatives of the function $P(x,t)$:

$$\frac{\partial P}{\partial x}(x,t) = \lim_{h \to 0} \frac{P(x+h,t) - P(x,t)}{h} \quad , \quad \frac{\partial P}{\partial t}(x,t) = \lim_{h \to 0} \frac{P(x,t+h) - P(x,t)}{h} .$$

Using that

$$P(x+h, t+h) \simeq P(t,x) + h\frac{\partial P}{\partial x}(x,t) + h\frac{\partial P}{\partial t}(x,t) ,$$

McKendrick obtained the following partial differential equation:

$$\frac{\partial P}{\partial t}(x,t) + \frac{\partial P}{\partial x}(x,t) + m(x)P(x,t) = 0 .$$

Such an equation appears naturally in population problems structured by a continuous variable, such as age in demography (see Chapter 25) or time since infection in epidemiology.

In 1921, William Ogilvy Kermack had been appointed in charge of the chemical section of the Royal College of Physicians Laboratory in Edinburgh. Kermack was born in 1898 in a small town in Scotland. He studied at Aberdeen University and started doing research in the field of organic chemistry in an industrial laboratory in Oxford. Despite becoming completely blind after an explosion in his Edinburgh laboratory in 1924, he continued his chemical work with the help of colleagues and students. Kermack also began to collaborate with McKendrick on the mathematical modelling of epidemics. Starting in 1927, they published together a series of "Contributions to the mathematical theory of epidemics" where they studied deterministic epidemic models. Let N be the population size with N large enough. Assume as in the 1926 article that people can be either susceptible, infected or recovered. If the disease is fatal then the third state is in fact death. Let $S(t)$, $I(t)$ and $R(t)$ be the number of people in each of the three states. The model is (in a simplified form) a system of three differential equations:

$$\frac{dS}{dt} = -aSI, \tag{16.2}$$

$$\frac{dI}{dt} = aSI - bI, \tag{16.3}$$

$$\frac{dR}{dt} = bI. \tag{16.4}$$

Hence, the number of new infections per unit of time is, as in the 1926 stochastic model, proportional to both the number of susceptible people and to the number of infected people. At the beginning of the epidemic, at time $t = 0$, a certain number of people are infected: $S(0) = N - I_0$, $I(0) = I_0$ and $R(0) = 0$, assuming $0 < I_0 < N$.

Although system (16.2)-(16.4) has no closed solution, several of its properties can be proved:

- the total population $S(t) + I(t) + R(t)$ stays constant and equal to N;
- $S(t)$, $I(t)$ and $R(t)$ stay nonnegative (as should be since these are populations);
- when $t \to +\infty$, $S(t)$ decreases to a limit $S_\infty > 0$, $I(t)$ tends to 0 and $R(t)$ increases to a limit $R_\infty < N$;
- moreover the formula

$$-\log \frac{S_\infty}{S(0)} = \frac{a}{b}(N - S_\infty), \tag{16.5}$$

gives implicitly S_∞ and therefore also the final epidemic size $R_\infty = N - S_\infty$.

Indeed, we see first that $\frac{d}{dt}(S+I+R) = 0$. So $S(t) + I(t) + R(t) = S(0) + I(0) + R(0) = N$. Equations (16.2) and (16.3) can be rewritten as

$$\frac{d}{dt}\left[S(t) e^{a\int_0^t I(\tau)\,d\tau}\right] = 0, \quad \frac{d}{dt}\left[I(t) e^{bt - a\int_0^t S(\tau)\,d\tau}\right] = 0.$$

It follows on one side that $S(t) = S(0) e^{-a\int_0^t I(\tau)\,d\tau} > 0$ and on the other side that $I(t) = I(0) e^{a\int_0^t S(\tau)\,d\tau - bt} > 0$. Equations (16.2) and (16.4) then show that the function $S(t)$ is decreasing and that the function $R(t)$ is increasing (in particular, $R(t) \geq 0$). Since $S(t) \geq 0$ and $R(t) \leq N$, the functions $S(t)$ and $R(t)$ do have limits when $t \to +\infty$. Since $I(t) = N - S(t) - R(t)$, $I(t)$ also has a limit when $t \to +\infty$, which can only be zero as can be seen by integrating (16.4). Equation (16.2) also shows that $-\frac{d}{dt}[\log S] = aI$. Integrating between $t = 0$ and $t = +\infty$, we find $\log S(0) - \log S_\infty = a\int_0^\infty I(t)\,dt$. Equation (16.3) can be rewritten as $\frac{dI}{dt} = -\frac{dS}{dt} - bI$. Integrating between $t = 0$ and $t = +\infty$, we get

$-I(0) = S(0) - S_\infty - b \int_0^\infty I(t)\,dt$. Combining the two results, we obtain formula (16.5), which shows that $S_\infty > 0$.

When the initial number of infected people I_0 is small compared with the population size N, which is often the case at the start of an epidemic in a city, formula (16.5) can be rewritten using $S_\infty = N - R_\infty$ as

$$-\log\left(1 - \frac{R_\infty}{N}\right) \simeq \mathscr{R}_0 \frac{R_\infty}{N}, \qquad (16.6)$$

where by definition

$$\mathscr{R}_0 = \frac{aN}{b}.$$

Equation (16.6) has a positive solution only if $\mathscr{R}_0 > 1$. So Kermack and McKendrick arrive at the following conclusion: the epidemic infects a non-negligible fraction of the population only if $\mathscr{R}_0 > 1$. There is a threshold for the population density $N^* = b/a$ below which epidemics cannot occur.

When the population size N is just above this threshold ($N = N^* + \varepsilon$), an epidemic of small amplitude happens. It follows from (16.6) that $R_\infty \simeq 2\varepsilon$. So $S_\infty \simeq N^* - \varepsilon$: *the epidemic brings the susceptible population as much below the threshold N^* as it was initially above.*

Indeed, using the approximation $-\log(1 - x) \simeq x + \frac{x^2}{2}$, equation (16.6) becomes

$$\frac{R_\infty}{N} + \frac{1}{2}\left(\frac{R_\infty}{N}\right)^2 \simeq \mathscr{R}_0 \frac{R_\infty}{N}.$$

So $R_\infty \simeq 2(\mathscr{R}_0 - 1)N = 2\frac{\varepsilon}{N^*}(N^* + \varepsilon) \simeq 2\varepsilon$.

As in Ross' malaria model (Chapter 12), the condition $\mathscr{R}_0 > 1$ has a simple interpretation. Since aN is the number of people that one infected person infects per unit of time at the beginning of the epidemic and since $1/b$ is the average infectious period, $\mathscr{R}_0 = aN/b$ is the average number of secondary cases due to one infected person at the beginning of the epidemic.

For fatal diseases, $R(t)$ is the cumulative number of deaths since the beginning of the epidemic and dR/dt is the number of deaths per unit of time. Kermack and McKendrick noticed that the graph of the function dR/dt in their mathematical model does have the bell shape that one expects from an epidemic curve (Fig. 16.4).

To draw dR/dt, they divided (16.2) by (16.4) to obtain $dS/dR = -aS/b$. So $S(t) = S(0)\exp(-aR(t)/b)$. Replacing this into equation (16.4) and using $S(t) + I(t) + R(t) = N$, they got the equation

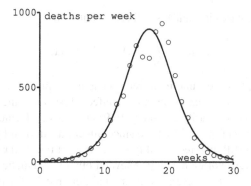

Fig. 16.4 The curve dR/dt as a function of time and the data for the number of deaths per week during a plague epidemic in Bombay in 1905-1906.

$$\frac{dR}{dt} = b\left[N - R - S(0)\exp\left(-\frac{a}{b}R\right)\right],\qquad(16.7)$$

which still cannot be solved explicitly. Nevertheless, if $\frac{a}{b}R(t)$ stays small during the entire epidemic, the approximation $\exp(-u) \simeq 1 - u + u^2/2$ gives

$$\frac{dR}{dt} \simeq b\left[N - R - S(0) + S(0)\frac{a}{b}R - S(0)\frac{a^2}{2b^2}R^2\right].\qquad(16.8)$$

This is a so-called Riccati equation with two constant solutions, one positive R_+ and one negative R_-, given by the roots of the second-order polynomial in R on the right-hand side of (16.8). Let $\tilde{R}(t)$ be the exact solution of (16.8) and set $Q(t) = \tilde{R}(t) - R_+$. Then $Q(t)$ satisfies a Bernoulli differential equation similar to those encountered by Daniel Bernoulli and Verhulst (see (4.5) and (6.1)). One can thus directly adapt formula (6.2) to get $Q(t)$. An easy but tedious computation shows that dQ/dt is of the form $\alpha/\cosh^2(\beta t - \gamma)$, where α, β and γ are constants that depend in a complicated way on the parameters of the model. As $dR/dt \simeq d\tilde{R}/dt = dQ/dt$, Kermack and McKendrick could choose (α, β, γ) to fit their data. Of course modern computers and software can easily solve numerically the differential equation (16.7) without going through these approximations.

The curve for dR/dt thus obtained fitted well the data for the number of deaths per week during the plague epidemic in Bombay between December 1905 and July 1906 (Fig. 16.4).

Kermack and McKendrick also considered the more general model where infectiousness $a(x)$ depends on the time x since infection and where the recovery rate $b(x)$ also depends on x. The equation giving the final epidemic size (when the initial number of infected cases is small) is still (16.6) but with

$$\mathscr{R}_0 = N \int_0^\infty a(x)\, e^{-\int_0^x b(y)\,dy}\, dx. \tag{16.9}$$

The parameter $\mathscr{R}_0$ has the same interpretation as in the previous case: it is the average number of secondary cases due to one infected person at the beginning of the epidemic. Notice the similarity between (16.9) and Lotka's formula (10.2) for $\mathscr{R}_0$ in demography: age is replaced by time since infection, survival by the probability $e^{-\int_0^x b(y)\,dy}$ of being still infected, fertility by the contact rate $N a(x)$.

Kermack and McKendrick developed several other mathematical models of epidemics during the 1930s. These are still the building blocks for most of the more complex models used nowadays in epidemiology. The parameter $\mathscr{R}_0$ –whose definition was generalized by Diekmann, Heesterbeek and Metz in 1990 –still plays a central role in the model's analysis.

McKendrick retired in 1941 and died in 1943. Between 1930 and 1933, Kermack coauthored a few articles on mathematical physics with William McCrea and Edmund Whittaker, both from the mathematics department at the University of Edinburgh. During the 1930s and 1940s, Kermack's team of chemists tried to synthesize new molecules with antimalarial activity, but with limited success. In 1938 Kermack coauthored with Philip Eggleton a popular book on elementary biochemistry, *The Stuff We're Made Of*. He was elected fellow of the Royal Society in 1944 and took the chair of biochemistry at Aberdeen University in 1949. He later served as the dean of the Faculty of Science. He retired in 1968 and died in 1970.

Further reading

1. Advisory Committee appointed by the Secretary of State for India, the Royal Society and the Lister Institute: Reports on plague investigations in India, XXII, Epidemiological observations in Bombay City. *J. Hyg.* **7**, 724–798 (1907). www.ncbi.nlm.nih.gov
2. Davidson, J.N., Yates, F., McCrea, W.H.: William Ogilvy Kermack 1898–1970. *Biog. Mem. Fellows R. Soc.* **17**, 399–429 (1971)
3. Diekmann, O., Heesterbeek, J.A.P., Metz, J.A.J.: On the definition and the computation of the basic reproduction ratio R_0 in models for infectious diseases in heterogeneous populations. *J. Math. Biol.* **28**, 365–382 (1990)
4. Gani, J.: A.G. McKendrick. In: Heyde, C.C., Seneta, E. (eds.) *Statisticians of the Centuries*, pp. 323–327. Springer, New York (2001). books.google.com
5. Harvey, W.F.: A.G. McKendrick 1876–1943. *Edinb. Med. J.* **50**, 500–506 (1943)
6. McKendrick, A.G.: Applications of mathematics to medical problems. *Proc. Edinb. Math. Soc.* **13**, 98–130 (1926)
7. Kermack, W.O., McKendrick, A.G.: A contribution to the mathematical theory of epidemics. *Proc. R. Soc. Lond. A* **115**, 700–721 (1927). gallica.bnf.fr

Chapter 17
Haldane and mutations (1927)

John Burdon Sanderson Haldane was born in 1892 in Oxford, where his father was professor of physiology at the university. Haldane studied at Eton College and after 1911 at New College of Oxford University. After focusing on mathematics in his first year, he turned to humanities. His studies were interrupted by the First World War, during which he served in France and Iraq. Having been wounded, he was sent as a military instructor to India. In 1915 he published a first article discussing genetic experiments on mice he had started before the war. In 1919 he became a fellow of New College, teaching physiology and studying respiration like his father. In 1923 he joined the biochemistry laboratory of F. G. Hopkins[1] at Cambridge University, where he focused on the kinetics of enzymes. He also published a science fiction novel, *Daedalus or Science and the Future* (1923), and an essay entitled *Callinicus, A Defense of Chemical Warfare* (1925). Between 1924 and 1934, he wrote a series of ten articles entitled "A mathematical theory of natural and artificial selection".

Fig. 17.1 Haldane (1892–1964)

---

[1] Frederick Gowland Hopkins, who received the Nobel Prize in Physiology or Medicine in 1929 for his work on vitamins.

N. Bacaër, *A Short History of Mathematical Population Dynamics*,
DOI 10.1007/978-0-85729-115-8_17, © Springer-Verlag London Limited 2011

In the fifth article of the series, published in 1927, Haldane reconsidered another genetic model that Fisher had studied in 1922, a model focusing on mutations. Fisher had studied the probability for a mutant gene to invade a population or to disappear. This problem is formally the same as that of Bienaymé, Galton and Watson concerning the extinction of family names. But Fisher made no reference to these works, though he may have read the article of Galton and Watson reproduced in the appendix of Galton's 1889 book *Natural Inheritance*. As in Chapter 9, call p_k the probability of a gene being transmitted to k offspring in the first generation ($k \geq 0$). Fisher considered also the generating function

$$f(x) = p_0 + p_1 x + p_2 x^2 + \cdots + p_k x^k + \cdots ,$$

except that he did not fix any upper bound for k: the sum can include an infinite number of terms. He realized that, starting from one individual with the mutant gene in generation 0, the probability of this gene being in k individuals is the coefficient of x^k in $f_1(x) = f(x)$ for generation 1, in $f_2(x) = f(f(x))$ for generation 2, in $f_3(x) = f(f(f(x)))$ for generation 3 etc. In this way, it becomes clear that the equation

$$f_n(x) = f(f_{n-1}(x)) \tag{17.1}$$

holds. This equation is much more practical than the equation $f_n(x) = f_{n-1}(f(x))$ derived by Watson. In particular, it follows from (17.1) that the extinction probability within n generations $x_n = f_n(0)$ satisfies the iteration formula $x_n = f(x_{n-1})$, as Bienaymé had already noticed.

As an example, Fisher considered the case of a plant with a mutant gene that can produce N seeds, each seed having a probability q of surviving to produce a new plant. The probability p_k of getting k offspring with the mutant gene is binomial:

$$p_k = \binom{N}{k} q^k (1-q)^{N-k}$$

for all $0 \leq k \leq N$ and $p_k = 0$ for $k > N$. The generating function is then $f(x) = (1 - q + qx)^N$. Let $\mathcal{R}_0 = Nq$ be the mean number of seeds that survive to produce a new plant. When N is large and q is small, then

$$f(x) = \left(1 + \frac{\mathcal{R}_0}{N}(x-1)\right)^N \simeq e^{\mathcal{R}_0(x-1)} = e^{-\mathcal{R}_0} \sum_{k=0}^{\infty} \frac{(\mathcal{R}_0 x)^k}{k!} .$$

The probability distribution (p_k) tends to $e^{-\mathcal{R}_0} (\mathcal{R}_0)^k / k!$, which is called a Poisson distribution. Fisher then computed the extinction probability within n generations, using $x_0 = 0$, $x_n \simeq e^{\mathcal{R}_0(x_{n-1}-1)}$ and the numerical values $N = 80$ and $q = 1/80$. In this case, $\mathcal{R}_0 = Nq = 1$. A tedious computation shows that $x_{100} \simeq 0.98$: *a mutant gene with no selective advantage* ($\mathcal{R}_0 = 1$) *disappears very slowly*. There is still a 2% chance for the gene to be present in the population after 100 generations. In 1922 Fisher did not push further the study of this model.

Continuing Fisher's work, Haldane first noticed in his 1927 article that, for any probability distribution (p_k) such that $p_0 > 0$, the equation $x = f(x)$ has exactly two roots in the interval $(0,1]$ when the mean number of offspring carrying the mutant gene $\mathscr{R}_0$ is strictly bigger than 1, i.e. when the mutant gene has a selective advantage. Moreover, the extinction probability x_∞, which is the limit of x_n as $n \to +\infty$, is the smallest of the two roots of $x = f(x)$: the gene has a nonzero probability of settling in the population. Unlike Bienaymé and Cournot, Haldane provided a proof for this conclusion.

Indeed, $f'(x) \geq 0$ and $f''(x) \geq 0$ on the interval $[0,1]$. In other words, the function $f(x)$ is nondecreasing and convex. The assumptions $f(0) = p_0 > 0$ and $f'(1) = \mathscr{R}_0 = p_1 + 2p_2 + 3p_3 + \cdots > 1$ imply that equation $f(x) = x$ has exactly two solutions in the interval $(0,1]$: $x = 1$ and x^* such that $0 < x^* < 1$. Haldane then referred to an article by Gabriel Koenigs from 1883, which showed that if $x_n = f(x_{n-1})$ and $x_n \to x_\infty$, then $x_\infty = f(x_\infty)$ and $|f'(x_\infty)| \leq 1$. When $f'(1) > 1$, the only possibility is that $x_\infty = x^*$.

For the case of a Poisson distribution with $f(x) = e^{\mathscr{R}_0(x-1)}$ and $\mathscr{R}_0$ just slightly bigger than 1, the extinction probability x_∞ is very close to 1. The equation $f(x_\infty) = x_\infty$ is then equivalent to

$$\mathscr{R}_0(x_\infty - 1) = \log x_\infty \simeq (x_\infty - 1) - \frac{(x_\infty - 1)^2}{2}.$$

It follows that $1 - x_\infty \simeq 2(\mathscr{R}_0 - 1)$. Haldane concluded that *the probability of the mutant gene not going extinct is twice its selective advantage* $\mathscr{R}_0 - 1$. Without citing Haldane, Fisher took as an example in his 1930 book the case where $\mathscr{R}_0 = 1.01$, which gives a 2% chance of the mutant gene not going extinct.

Haldane became a fellow of the Royal Society in 1932. He left Cambridge to become professor of genetics and later biometry at University College in London. He was then particularly interested in human genetics: estimation of mutation rates, genetic maps of chromosomes etc. Beside his scientific books (*Animal Biology* in 1927 with Julian Huxley, *Enzymes* in 1930 and *The Causes of Evolution* in 1932, *The Biochemistry of Genetics* in 1954), he published a large number of articles on science in the press (for example, on the origin of life) and some essays (*The Inequality of Man* in 1932, *The Philosophy of a Biologist* in 1935, *The Marxist Philosophy and the Sciences* in 1938, *Heredity and Politics* in 1938 and *Science Advances* in 1947). After several visits to Spain during the civil war, he tried to convince his own country to build shelters against air bombing. During the Second World War, he worked on respiration problems in submarines. A member of the communist party since 1942, he resigned in 1950 because of the official rejection of Mendelian genetics in the USSR due to the influence of Lysenko. In 1957 he settled in India, where he continued his research, first at the Indian Statistical Institute in Calcutta and later in Bhubaneswar. Having become an Indian citizen, he died in 1964.

Further reading

1. Clark, R.: *J.B.S., The Life and Work of J.B.S. Haldane*. Hodder and Stoughton, London (1968)
2. Haldane, J.B.S.: A mathematical theory of natural and artificial selection, Part V, Selection and mutation. *Proc. Camb. Philos. Soc.* **23**, 838–844 (1927)
3. Haldane, J.B.S.: *The Causes of Evolution*. Longmans (1932) Reprint, Princeton University Press (1990). books.google.com
4. Pirie, N.W.: John Burdon Sanderson Haldane 1892-1964. *Biog. Mem. Fellows R. Soc.* **12**, 218–249 (1966)

Chapter 18
Erlang and Steffensen on the extinction problem (1929–1933)

Agner Krarup Erlang was born in 1878 in Lønborg, Denmark. His father was a schoolmaster. Between 1896 and 1901, the young Erlang studied mathematics, physics and chemistry at the University of Copenhagen. He then taught several years in high schools while keeping an interest in mathematics, especially probability theory. He met Jensen, chief engineer at the Copenhagen Telephone Company and an amateur mathematician, who convinced him in 1908 to join the new research laboratory of the company. Erlang started to publish articles on the applications of probability theory to the management of telephone calls. In 1917 he discovered a formula for waiting times, which was quickly used by telephone companies throughout the world. His articles, first published in Danish, were then translated in several other languages.

Fig. 18.1 Erlang (1878–1929)

In 1929 Erlang became interested in the same problem of extinction that Bienaymé, Galton and Watson had studied before him for family names and that Fisher and Haldane had studied for mutant genes. Like his predecessors, he was not aware

N. Bacaër, *A Short History of Mathematical Population Dynamics*,
DOI 10.1007/978-0-85729-115-8_18, © Springer-Verlag London Limited 2011

of all the works that had been published. Calling again p_k the probability for one individual to have k offspring, he noticed that the probability x_n of extinction within n generations satisfies

$$x_n = p_0 + p_1 x_{n-1} + p_2 x_{n-1}^2 + \cdots = f(x_{n-1})$$

with $x_0 = 0$. He noticed also that the overall extinction probability x_∞, which is the limit of x_n as $n \to +\infty$, is a solution of the equation $x_\infty = f(x_\infty)$. He realized that $x = 1$ was always a solution and that another solution existed between 0 and 1 when the average number of offspring $\mathscr{R}_0 = f'(1)$ is bigger than 1. But it seems that he could not figure out which of these two solutions was the right one. Like Galton, he submitted the problem in 1929 to a Danish mathematics journal, *Matematisk Tidsskrift*:

> Question 15. When the probability that an individual has k children is p_k, where $p_0 + p_1 + p_2 + \cdots = 1$, find the probability that his family dies out.

Unfortunately, Erlang died that same year 1929 at the age of 51. As a matter of fact, he died childless[1].

A professor of actuarial mathematics at the University of Copenhagen, Johan Frederik Steffensen, took up Erlang's question. He published in 1930 his solution in the same Danish journal: the probability of extinction x_∞ is always the smallest root of the equation $x = f(x)$ in the closed interval $[0, 1]$, as Bienaymé and Haldane had already noticed. Steffensen's proof is the one to be found in modern textbooks.

> Indeed, we saw that the extinction probability x_∞ is a solution of $x = f(x)$ in the closed interval $[0, 1]$. Let x^* be the smallest such solution. By definition, $x^* \leq x_\infty$. Steffensen noticed first that $x^* = f(x^*) \geq p_0 = x_1$. Assume by induction that $x^* \geq x_n$. Then $x^* = f(x^*) \geq f(x_n) = x_{n+1}$ since the function $f(x)$ is increasing. So $x^* \geq x_n$ for all n. Taking the limit, $x^* \geq x_\infty$. So $x_\infty = x^*$. Q.E.D.

Steffensen gave also a more formal explanation as to why $x = 1$ is the only root of $x = f(x)$ when the mean number of offspring $\mathscr{R}_0 = f'(1)$ is smaller or equal to 1 (Fig. 18.2a) and why there is only one other root different from $x = 1$ in the case where $\mathscr{R}_0 > 1$ (Fig. 18.2b). Notice that $\mathscr{R}_0 = f'(1)$ is the slope of the function $f(x)$ at $x = 1$.

> He noticed that for any root of $x = f(x)$,
>
> $$1 - x = 1 - f(x) = 1 - p_0 - \sum_{k=1}^{+\infty} p_k x^k = \sum_{k=1}^{+\infty} p_k (1 - x^k).$$

---

[1] In his memory, the International Telephone Consultative Committee decided in 1946 to call "erlang" the unit of measure of the intensity of telephone traffic. "Erlang" is also the name given to a programming language by the company Ericsson.

Assuming $x \neq 1$ and dividing by $1 - x$, we get

$$1 = p_1 + p_2(1+x) + p_3(1+x+x^2) + \cdots . \tag{18.1}$$

When x increases from 0 to 1, the right-hand side of Eq. (18.1) increases from $1 - p_0$ to $\mathcal{R}_0 = f'(1)$. If $\mathcal{R}_0 < 1$, then Eq. (18.1) has no solution. If $\mathcal{R}_0 \geq 1$ and if we exlude the trivial case where $p_1 = 1$, then the right-hand side of Eq. (18.1) is a strictly increasing function of x. Otherwise there would be no $k \geq 2$ such that $p_k \neq 0$ and $\mathcal{R}_0$ would be equal to $p_1 < 1$. In conclusion, (18.1) has one and only one solution in the interval $[0, 1]$ when $\mathcal{R}_0 \geq 1$.

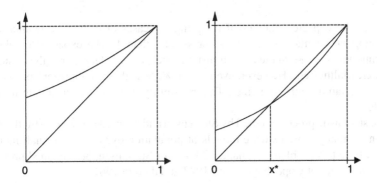

Fig. 18.2 Graph of the functions $y = x$ and $y = f(x)$ in the example of Chapter 17, $f(x) = e^{\mathcal{R}_0(x-1)}$, with $\mathcal{R}_0 = 0.75 < 1$ (left) or $\mathcal{R}_0 = 1.5 > 1$ (right).

Steffensen, who was also president of the Danish Actuarial Society and of the Danish Mathematical Society, was invited to the University of London in 1930. His British colleague W. P. Elderton told him about the work of Galton and Watson. In 1933 Steffensen published a new article in the annals of the Institut Henri Poincaré, where he had given a conference in 1931. He summarized the results of his article in Danish and compared them with those of Watson. He also showed that the mathematical expectation of the number of offspring in generation n is equal to $(\mathcal{R}_0)^n$.

Indeed, let $p_{k,n}$ be the probability that there are k offspring in generation n, starting from one individual in generation 0. In his 1930 article, Steffensen had noticed like his predecessors that the generating function

$$f_n(x) = \sum_{k=0}^{+\infty} p_{k,n} x^k$$

relative to generation n satisfies $f_1(x) = f(x)$ and

$$f_n(x) = f(f_{n-1}(x)).$$ (18.2)

Let M_n be the expectation of the number of offspring in generation n. Then

$$M_n = \sum_{k=1}^{+\infty} k\, p_{k,n} = f_n'(1).$$

Deriving formula (18.2), we get $f_n'(x) = f'(f_{n-1}(x)) \times f_{n-1}'(x)$. So $M_n = f_n'(1) = f'(f_{n-1}(1)) \times f_{n-1}'(1) = f'(1) \times M_{n-1} = \mathcal{R}_0 \times M_{n-1}$. Since $M_1 = f_1'(1) = f'(1) = \mathcal{R}_0$, it follows that $M_n = (\mathcal{R}_0)^n$ for all n.

Hence the expected number of offspring increases or decreases geometrically depending on whether $\mathcal{R}_0$ is bigger or smaller than 1. The expected number of offspring behaves like in the deterministic models of population growth considered by Euler, Malthus etc. However, even when $\mathcal{R}_0 > 1$, there is a nonzero probability x_∞ that the family will go extinct. This possibility does not occur in deterministic models.

The stochastic process studied by Steffensen and his predecessors is still the basic element of many more realistic models of population dynamics. We shall mention one last time this problem in Chapter 20. As for Steffensen, he remained professor at the University of Copenhagen until 1943 and died in 1961.

Further reading

1. Brockmeyer, E., Halstrøm, H.L., Jensen, A.: The life and works of A.K. Erlang. *Trans. Dan. Acad. Techn. Sci.* **2** (1948). oldwww.com.dtu.dk
2. Erlang, A.K.: Opgave Nr. 15. *Mat. Tidsskr. B*, 36 (1929) Translation in Guttorp (1995)
3. Guttorp, P.: Three papers on the history of branching processes. *Int. Stat. Rev.* **63**, 233–245 (1995). www.stat.washington.edu
4. Heyde, C.C.: Agner Krarup Erlang. In: Heyde, C.C., Seneta, E. (eds.) *Statisticians of the Centuries*, pp. 328–330. Springer, New York (2001)
5. Ogborn, M.E.: Johan Frederik Steffensen, 1873–1961. *J. R. Stat. Soc. Ser. A* **125**, 672–673 (1962)
6. Steffensen, J.F.: Om Sandssynligheden for at Afkommet uddør. *Mat. Tidsskr. B*, 19–23 (1930) English translation in Guttorp (1995)
7. Steffensen, J.F.: Deux problèmes du calcul des probabilités. *Ann. Inst. Henri Poincaré* **3**, 319–344 (1933). archive.numdam.org

Chapter 19
Wright and random genetic drift (1931)

Sewall Wright was born in Massachusetts in 1889. He did his undergraduate studies in a small college in Illinois where his father taught economics. After a master's degree in biology from the University of Illinois at Urbana and a summer school at Cold Spring Harbor Laboratory, Wright did a PhD at Harvard University on the inheritance of coat colour in the guinea pig. Between 1915 and 1925, he continued to work on inbreeding experiments with guinea pigs at the Animal Husbandry Division of the United States Department of Agriculture in Washington. He developed the "method of path coefficients" to analyze these experiments. He then joined the department of zoology at the University of Chicago.

Fig. 19.1 Wright
(1889–1988)

Influenced by Fisher's 1922 article on population genetics (see Chapter 14), Wright wrote in 1925 a long article entitled *Evolution in Mendelian populations*, which was finally published in 1931. He studied in particular a mathematical model that appeared also implicitly in Fisher's 1930 book on *The Genetical Theory of Natural Selection*. As in the Hardy–Weinberg law, this model considers the case where there are just two possible alleles A and a for one locus, but the population is not assumed to be infinitely large. The point is to see if removing this assumption has

N. Bacaër, *A Short History of Mathematical Population Dynamics*,
DOI 10.1007/978-0-85729-115-8_19, © Springer-Verlag London Limited 2011

some influence on the genetic composition of the population. So let N be the total number of individuals, which is assumed to be the same in all generations. Each individual has two alleles. So there is a total of $2N$ alleles in the population in each generation. The model assumes also that mating occurs at random. If there are i alleles A and $2N - i$ alleles a in generation n, then an allele chosen at random among individuals in generation $n + 1$ will be A with a probability $\frac{i}{2N}$ and a with a probability $1 - \frac{i}{2N}$. The number of A alleles in generation $n + 1$ will therefore be equal to j with a probability

$$p_{i,j} = \binom{2N}{j} \left(\frac{i}{2N}\right)^j \left(1 - \frac{i}{2N}\right)^{2N-j}, \tag{19.1}$$

where $\binom{2N}{j}$ is the binomial coefficient equal to $(2N)!/j!/(2N - j)!$. Let X_n be the number of A alleles in generation n: it is a random variable (Fig. 19.2). One can

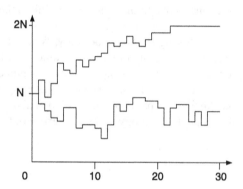

Fig. 19.2 Two simulations showing the variations of the number X_n of A alleles during 30 generations if $N = 20$ and $X_0 = 10$.

show that the expectation of X_{n+1} knowing that $X_n = i$ is equal to i: this is reminiscent of the Hardy–Weinberg law, where the frequency of allele A remained constant through generations.

Indeed, consider the generating function

$$f(x) = \sum_{j=0}^{2N} p_{ij} x^j = \left(1 - \frac{i}{2N} + \frac{ix}{2N}\right)^{2N},$$

The expectation of X_{n+1} knowing that $X_n = i$ is then

$$\sum_{j=0}^{2N} j p_{i,j} = f'(1) = i. \tag{19.2}$$

However it is possible in this model that, starting from an initial condition $X_0 = i$ with $0 < i < 2N$, the event $X_n = 0$ occurs by chance after a certain number of generations. In such a case, all alleles would be of type a and X_n would remain equal to 0 in all future generations. The same fixation would happen with allele A if $X_n = 2N$ after a certain number of generations. In summary, when the population is assumed infinitely large as in the Hardy–Weinberg model, the two alleles cannot disappear because their frequencies remain constant. When one takes into account the finite size of populations, as in the Fisher–Wright model, the frequencies of the two alleles fluctuate and one of the alleles can (and will) disappear.

Starting from $X_0 = i$, one can easily compute the probability Q_i for the population to be fixed in state $X = 0$. Indeed, Q_i has to satisfy the "boundary conditions"

$$Q_0 = 1, \quad Q_{2N} = 0. \tag{19.3}$$

Moreover,

$$Q_i = \sum_{j=0}^{2N} p_{i,j} Q_j, \tag{19.4}$$

because $p_{i,j} Q_j$ is the probability of being fixed in state $X = 0$ starting from $X_0 = i$ and passing through $X_1 = j$. Since $\sum_j p_{i,j} = 1$, we see using (19.2) that $Q_i = 1 - \frac{i}{2N}$ is the solution of system (19.3)-(19.4). Hence the probability that, starting from i alleles of type A in a population of size N, the system evolves towards a population containing only the allele a is equal to $1 - \frac{i}{2N}$. Similarly, the probability that it evolves towards a population containing only the allele A is equal to $\frac{i}{2N}$.

Wright managed to show that the number of generations that elapse before fixation in one of the two extreme states is of the order of $2N$ generations (Fig. 19.3). For populations of several millions of individuals, this time would be so long that the frequencies of the alleles could be considered as almost constant, as in the Hardy–Weinberg law.

Indeed, assume that there are i_0 alleles of type A in the population in generation 0. Let $u_i^{(n)}$ be the probability that there are i alleles of type A in the population in generation n. Then

$$u_j^{(n+1)} = \sum_{i=0}^{2N} u_i^{(n)} p_{i,j}$$

for all $j = 0, \ldots, 2N$. We have already seen that, when $n \to +\infty$,

$$u_0^{(n)} \to 1 - \frac{i_0}{2N}, \quad u_{2N}^{(n)} \to \frac{i_0}{2N}, \quad u_i^{(n)} \to 0$$

for all $0 < i < 2N$. Wright noticed that if $u_i^{(n)} = v$ for all $i = 1, \ldots, 2N - 1$, then

$$u_j^{(n+1)} = v \binom{2N}{j} \sum_{i=1}^{2N-1} \left(\frac{i}{2N}\right)^j \left(1-\frac{i}{2N}\right)^{2N-j} \tag{19.5}$$

for all $1 < j < 2N$ because $p_{0,j} = p_{2N,j} = 0$. When N is large enough,

$$\frac{1}{2N} \sum_{i=1}^{2N-1} \left(\frac{i}{2N}\right)^j \left(1-\frac{i}{2N}\right)^{2N-j} \simeq \int_0^1 x^j (1-x)^{2N-j}\, dx$$

$$= \frac{j!\,(2N-j)!}{(2N+1)!}, \tag{19.6}$$

the value of the integral being obtained by successive integrations by parts. Combining (19.5) and (19.6), we arrive finally for $0 < j < 2N$ at

$$u_j^{(n+1)} \simeq \frac{2N}{2N+1} v = \left(1 - \frac{1}{2N+1}\right) u_j^{(n)}.$$

So the probabilities $u_j^{(n)}$ for all $0 < j < 2N$ decrease at a rate of about $1/2N$ per generation. This rate is very slow if N is large. There is almost no decrease if, for example, N is of the order of magnitude of millions.

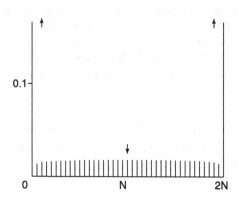

Fig. 19.3 Probability that there are i alleles A in the population ($i = 0, \ldots, 2N$ on the horizontal axis) after 30 generations if $N = 20$ and $X_0 = 10$.

In 1922 Fisher had already tried to estimate this rate of fixation ($1/2N$) but had missed a factor 2. In any case, the two scientists disagreed over the typical size N of breeding populations. For the theory of evolution, Wright's work suggested that random genetic drift in a small population could be a mechanism for the origin of species. Biologists working on the classification of species had indeed noticed that differences between species or subspecies often had no apparent explanation in terms of natural selection. This idea was strongly opposed during the 1940s and 1950s by Fisher and his colleague E. B. Ford who both thought that random genetic drift was negligible compared to natural selection. They referred in particular to their study of the fluctuations of gene frequencies in a small isolated population of moths (*Panaxia dominula*) near Oxford, where the three genotypes for a certain gene (common homozygote, heterozygote and rare homozygote) could be distinguished by sight. Another famous controversy over the respective influence of natural selection and of random drift focused on snails of the genus *Cepaea*. More realistic models of evolution now combine random drift, selection, mutation, migration, nonrandom mating etc. The role of random drift was later reemphasized by the Japanese scientist Motoo Kimura with his "neutral theory of molecular evolution". Another outgrowth was the development of coalescent theory (introduced by John Kingman in 1982), which traces the ancestry of genes backward in time back to the point where they have a single common ancestor.

Wright became a member of the National Academy of Sciences in 1934. He worked for many years with Theodosius Dobzhansky on the genetics of natural populations of flies (*Drosophila pseudoobscura*) in the Death Valley region. He retired from the University of Chicago in 1955 but continued another five years as a professor at the University of Wisconsin-Madison. Between 1968 and 1978, he published a four-volume treatise summarizing his work on *Evolution and the Genetics of Populations*. He received the Balzan Prize in 1984 and died in 1988 at the age of 98.

Further reading

1. Fisher, R.A.: *The Genetical Theory of Natural Selection*. Clarendon Press, Oxford (1930). www.archive.org
2. Hill, W.G.: Sewall Wright, 21 December 1889–3 March 1988. *Biog. Mem. Fellows R. Soc.* **36**, 568–579 (1990)
3. Kimura, M.: *The Neutral Theory of Molecular Evolution*. Cambridge University Press (1983). books.google.com
4. Provine, W.B.: *Sewall Wright and Evolutionary Biology*. University of Chicago Press (1989). books.google.com
5. Wright, S.: Evolution in Mendelian populations. *Genetics* **16**, 97–159 (1931). www.esp.org
6. Wright, S.: *Evolution and the Genetics of Populations*, Vol. 2, Theory of Gene Frequencies. University of Chicago Press (1969). books.google.com

Chapter 20
The diffusion of genes (1937)

In 1937, two articles were published introducing a new approach to the study of spatial heterogeneity in population dynamics. Fisher was the author of the first article, entitled *The wave of advance of advantageous genes*, which appeared in the *Annals of Eugenics*. He studied the spatial propagation of a favourable gene in a population. As a simplification, he considered a space reduced to just one dimension and called $u(x,t)$ the proportion of the population located at point x at time t that possesses the favourable gene. So $0 \leq u(x,t) \leq 1$. To include natural selection, he used equation (14.6) with a continuous time variable

$$\frac{\partial u}{\partial t} = au(1-u),$$

where a is a positive parameter. For a given value of x, we recognize Verhulst's logistic equation (see Chapter 6) with a solution $u(x,t)$ that tends to 1 as $t \to +\infty$. Furthermore, Fisher assumed that the offspring of an individual located at point x with the favourable gene do not stay at the same point but disperse randomly in the neighbourhood of x. By analogy with physics, he argued that one must add a diffusion term to the equation for $u(x,t)$, leading to the partial differential equation

$$\frac{\partial u}{\partial t} = au(1-u) + D\frac{\partial^2 u}{\partial x^2}. \tag{20.1}$$

When the selection coefficient a is zero, this reduces to the diffusion equation introduced by Fourier in his theory of heat and later used by Fick for the diffusion of physical particles. In 1904, Ronald Ross had started considering random dispersal in population dynamics. He was then wondering how the density of mosquitoes decreases as the distance from a breeding site increases. The problem had come to the attention of Karl Pearson and Lord Rayleigh. By 1937 the body of scientific literature on diffusion equations had grown considerably, in particular following Einstein's work on Brownian motion.

Fisher showed that there exist solutions of equation (20.1) of the form $u(x,t) = U(x+vt)$ satisfying the three conditions

$$0 \leq u(x,t) \leq 1, \quad u(x,t) \xrightarrow[x \to -\infty]{} 0, \quad u(x,t) \xrightarrow[x \to +\infty]{} 1,$$

provided that $v \geq v^*$ where

$$v^* = 2\sqrt{aD}.$$

These solutions connect the steady state $u = 1$ with the favourable gene to the steady state $u = 0$ with no such gene. They represent waves propagating at speed v in the direction of decreasing values of x. Indeed, $u(x - vT, t + T) = u(x,t)$: the part of the wave that was in position x at time t moves to position $x - vT$ at time $t + T$.

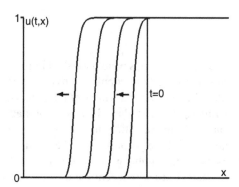

Fig. 20.1 Propagation from left to right of a favourable gene at the speed v^*. The gene frequency $u(t,x)$ at $t = 0$ is a step function.

Indeed, setting $z = x + vt$, Fisher noticed that if $u(x,t) = U(z)$, then $\frac{\partial u}{\partial t} = vU'(z)$, $\frac{\partial u}{\partial x} = U'(z)$ and $\frac{\partial^2 u}{\partial x^2} = U''(z)$. If u is a solution of equation (20.1), then

$$vU'(z) = aU(z)\left(1 - U(z)\right) + DU''(z). \tag{20.2}$$

When u is close to 0, i.e., when $z \to -\infty$, Fisher expected that $U(z) \to 0$ and $U'(z) \to 0$. Calling k the limit of $U'(z)/U(z)$ when $z \to -\infty$, we know from L'Hôpital's rule that $U''(z)/U'(z)$ also tends to k. Therefore, $U''(z)/U(z) = [U''(z)/U'(z)] \times [U'(z)/U(z)]$ tends to k^2. Dividing equation (20.2) by $U(z)$ and letting z tend to $-\infty$, we arrive at a second-order equation $Dk^2 - vk + a = 0$. But k must be a real number. So the discriminant of this equation has to be nonnegative: $v^2 - 4aD \geq 0$, or $v \geq 2\sqrt{aD} = v^*$. Hence, $v \geq v^*$ is a necessary condition for the existence of a wave propagating at the speed v. It is also a sufficient condition, as explained below.

Fisher noticed that only the wave which propagates exactly at the speed v^* is selected for a large class of initial conditions, e.g., for the step function: $u(x,0) = 0$ for $x < 0$, $u(x,0) = 1$ for $x \geq 0$. Figure 20.1 shows how this discontinuous initial condition becomes progressively a smooth wave propagating in the direction of decreasing x at the speed v^*.

That same year 1937, and independently of Fisher's work, Andrey Nikolaevich Kolmogorov, Ivan Georgievich Petrovsky and Nikolay Semenovich Piskunov studied the same problem of propagation of a dominant gene.

Kolmogorov was born in 1903 in Tambov, Russia. During his mathematics studies at Moscow State University, he did some important work on trigonometric series. He became a researcher at the Mathematics and Mechanics Institute in 1929 and a university professor in 1931. He worked on stochastic processes and their link with differential and partial differential equations. In 1933 he published a treatise laying the modern foundations of probability theory. His research interests included topology, approximation theory, Markov chains, Brownian motion and also applications to biological problems. In 1935 he published an article on genetics discussing the results of Hardy, Fisher and Wright. In 1936 he published an article on a generalization of the Lotka–Volterra system.

Fig. 20.2 Kolmogorov (1903–1987) and Petrovsky (1901–1973)

Petrovsky was born in 1901 in Sevsk. He also studied mathematics at Moscow State University, where he became a professor in 1933. He worked mainly on the theory of partial differential equations and on the topology of real algebraic curves, but also wrote some articles on ordinary differential equations and on the theory of probability. Piskunov, who was born in 1908, was another former mathematics student at Moscow State University.

During the 1930s Kolmogorov had contacts with A. S. Serebrovsky, a pioneer of population genetics in Moscow. It was then becoming increasingly dangerous to defend Mendelian genetics in the USSR because of the rise of Lysenko, an agronomist who had managed to convince Stalin that Mendelian genetics was mere "bourgeois pseudoscience". The Seventh International Congress of Genetics, originally scheduled for 1937 in Moscow, was cancelled. Many Soviet geneticists were executed or sent to labour camps.

In their 1937 article entitled "A study of the diffusion equation with increase in the amount of substance and its application to a biological problem", which was published in the *Bulletin of Moscow State University*, Kolmogorov, Petrovsky and Piskunov nevertheless used a mathematical model based on Mendelian genetics. Their model was a partial differential equation of the form

$$\frac{\partial u}{\partial t} = f(u) + D\frac{\partial^2 u}{\partial x^2} \tag{20.3}$$

where $u(x,t)$ is again the frequency of the favourable gene at point x and time t. The function $f(u)$ is assumed to satisfy several conditions: $f(0) = f(1) = 0$, $f(u) > 0$ if $0 < u < 1$, $f'(0) > 0$ and $f'(u) < f'(0)$ if $0 < u \leq 1$. The authors showed a result that is analogous to that of Fisher but with a more rigorous proof: if the initial condition is such that $0 \leq u(x,0) \leq 1$, $u(x,0) = 0$ for all $x < x_1$ and $u(x,0) = 1$ for all $x > x_2 \geq x_1$, then the gene propagates at the speed $v^* = 2\sqrt{f'(0)D}$.

Looking for a solution $u(x,t) = U(z)$ where $z = x + vt$ leads to the obvious generalization of equation (20.2) $vU'(z) = f(U(z)) + DU''(z)$. This second-order differential equation can be rewritten as a system of first-order differential equations

$$\frac{dU}{dz} = p, \quad \frac{dp}{dz} = \frac{vp - f(U)}{D}. \tag{20.4}$$

Recall that $U(z)$ should be such that $U(z) \to 0$ as $z \to -\infty$ and $U(z) \to 1$ as $z \to +\infty$. Near the steady state $(U = 0, p = 0)$ of system (20.4), we have $f(U) \simeq f'(0)U$. So (20.4) can be approximated by the linear system

$$\frac{dU}{dz} = p, \quad \frac{dp}{dz} = \frac{vp - f'(0)U}{D}. \tag{20.5}$$

Looking for exponential solutions of the form $U(z) = U_0 e^{kz}$ and $p(z) = p_0 e^{kz}$ yields the characteristic equation $Dk^2 - vk + f'(0) = 0$, as in Fisher's article. Again k must be real (otherwise u would oscillate and take negative values). Thus $v \geq 2\sqrt{f'(0)D} = v^*$. The two roots for k are then real and positive. If $v > v^*$, the two roots are different and the steady state $(U = 0, p = 0)$ is an unstable node. If $v = v^*$, the two roots are identical and $(U = 0, p = 0)$ is an unstable degenerate node as shown in Fig. 20.3. Similarly, system (20.4) near the steady state $(U = 1, p = 0)$ leads to the linear system

$$\frac{d(U-1)}{dz} = p, \quad \frac{dp}{dz} = \frac{vp - f'(1)(U-1)}{D}$$

and to the characteristic equation $Dk^2 - vk + f'(1) = 0$. The discriminant is $v^2 - 4Df'(1) \geq 0$ since $f'(1) \leq 0$. If $f'(1) < 0$, there are two real roots of opposite sign and $(U = 1, p = 0)$ is a saddle point. If $f'(1) = 0$, one root is zero and the other one is positive (see Fig. 20.3). A detailed analysis shows in fact that for all $v \geq 2\sqrt{f'(0)D}$ there is a unique integral curve joining the two steady states $(U = 0, p = 0)$ and $(U = 1, p = 0)$, as in the special case of Fig. 20.3.

Kolmogorov, Petrovsky and Piskunov went on to show rigorously that the partial differential equation (20.3) has a unique solution $u(x,t)$ satisfying the initial condition, that this solution is such that $0 < u(x,t) \leq 1$ for all x and $t > 0$, that $u(x,t)$ remains an increasing function of x if it is so at $t = 0$ and finally that $u(x,t)$ does converge towards a wave profile propagating at the speed v^*. The proofs are too long to be summarized here.

Notice that the function $f(u) = au(1-u)$ used by Fisher does satisfy all these conditions with $f'(0) = a$. Inspired by equation (14.5), Kolmogorov, Petrovsky and Piskunov considered the function $f(u) = au(1-u)^2$, which satisfies the same conditions and gives the same propagation speed.

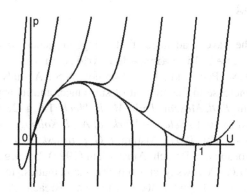

Fig. 20.3 Diagram (U,p) showing some integral curves of system (20.5) and in particular the unique curve joining $(U = 1, p = 0)$ to $(U = 0, p = 0)$, which is the one giving the shape of the propagating wave. Here, $f(u) = au(1-u)^2$, $a = 1$, $D = 1$ and $v = v^* = 2$.

The articles by Fisher and by Kolmogorov, Petrovsky and Piskunov were the starting point for the construction of many mathematical models with geographic diffusion in genetics, ecology and epidemiology. These models are known as "reaction-diffusion systems".

As for Kolmogorov, starting in 1938 he also studied the problem of extinction of family names considered by Bienaymé, Galton, Watson, Fisher, Haldane, Erlang and Steffensen: he called the stochastic process that is common to all these works the "branching process". In 1939 he became member of the USSR Academy of

Sciences. He later made important contributions to the problem of turbulence in fluid mechanics (1941), to the theory of dynamical systems linked with celestial mechanics (1953) and to information theory (starting in 1956). He also contributed to the writing of an encyclopedia and of high school and university textbooks, helped establish an experimental high school and edited a popular science magazine. He received many international prizes (including the Balzan prize in 1963 and the Wolf prize in 1980) and died in Moscow in 1987.

Petrovsky became the dean of the Mechanics and Mathematics Faculty of Moscow State University in 1940. He was the rector of the university from 1951 until his death in 1973. He was a full member of the USSR Academy of Sciences from 1946 and the president of the International Congress of Mathematicians that was held in Moscow in 1966. He also wrote textbooks on ordinary differential equations, partial differential equations and integral equations. Piskunov became a professor at a military academy. His textbook on differential and integral calculus was used by many technical universities. He died in 1977.

Further reading

1. Fisher, R.A.: The wave of advance of advantageous genes. *Ann. Eugen.* **7**, 355–369 (1937). digital.library.adelaide.edu.au
2. Kolmogorov, A.N., Petrovskii, I.G., Piskunov, N.S.: A study of the diffusion equation with increase in the amount of substance, and its application to a biological problem. *Bull. Moscow Univ. Math. Mech.* **1**:6, 1–26 (1937) Reprinted in: V.M. Tikhomirov (ed.) *Selected Works of A. N. Kolmogorov*, vol. 1, p. 242–270. Kluwer, Dordrecht (1991). Also in: *I. G. Petrowsky Selected Works*, Part II, pp. 106–132. Gordon and Breach, Amsterdam (1996). books.google.com
3. Oleinik, O.A.: I.G. Petrowsky and modern mathematics. In: *I. G. Petrowsky Selected Works*, Part I, pp. 4–30. Gordon and Breach, Amsterdam (1996). books.google.com
4. Pearson, K.: *Mathematical Contributions to the Theory of Evolution, XV, A Mathematical Theory of Random Migration.* Dulau, London (1906). www.archive.org
5. Rosenfeld, B.A.: Reminiscences of Soviet Mathematicians. In: Zdravkovska, S., Duren, P.L. (eds.) *Golden Years of Moscow Mathematics*, 2nd edn., pp. 75–100. American Mathematical Society (2007). books.google.com
6. Shiryaev, A.N. (ed.): *Selected Works of A. N. Kolmogorov*, vol. 2. Kluwer, Dordrecht (1992). books.google.com
7. Shiryaev, A.N.: Andrei Nikolaevich Kolmogorov (April 25, 1903 to October 20, 1987). In: *Kolmogorov in Perspective*, pp. 1–88. American Mathematical Society (2000). books.google.com

Chapter 21
The Leslie matrix (1945)

Patrick Holt Leslie was born in 1900 near Edinburgh in Scotland. He studied at Christ Church College of Oxford University and obtained in 1921 a bachelor's degree in physiology. But he could not finish his medical studies because of health problems. After a few years working as an assistant in bacteriology in the department of pathology, he turned to statistics and joined in 1935 the Bureau of Animal Population, a new research centre set up by Charles Elton. The purpose of this centre was to study the fluctuations of animal populations through field studies and laboratory experiments. Most of the research was done on rodents: analysis of the cycles of the hare and its predator the lynx using the archives of the Hudson Bay Company in Canada, follow-up of the territorial expansion of the grey squirrel at the expense of the red squirrel in England, data collection on voles in the neighbourhood of Oxford and so on. Leslie applied to the data on voles the methods developed by Lotka for human demography. During the Second World War, the centre's research focused on control methods of rats and mice in silos.

Fig. 21.1 P. H. Leslie (1900–1972)

In 1945 Leslie published his most famous article in *Biometrika*, a journal which had been founded by Galton, Pearson and Weldon in 1901. The article was entitled *On the use of matrices in certain population mathematics*. Leslie considered a model

N. Bacaër, *A Short History of Mathematical Population Dynamics*, DOI 10.1007/978-0-85729-115-8_21, © Springer-Verlag London Limited 2011

for the growth of the number of females in an animal population, e.g. a population of rats (but it could also be humans). The population is divided in $K+1$ age groups: $P_{k,n}$ is the number of females aged k at time n ($k = 0, 1, \ldots, K$; $n = 0, 1, \ldots$). Call f_k the fertility at age k or more precisely the number of daughters born per female between time n and time $n+1$. Then K is the maximum age with nonzero fertility ($f_K > 0$). Call s_k the probability for an animal aged k to survive at least until age $k+1$. Then the population's age structure is given by the following set of equations:

$$\begin{cases} P_{0,n+1} = f_0 P_{0,n} + f_1 P_{1,n} + \cdots + f_K P_{K,n} \\ P_{1,n+1} = s_0 P_{0,n} \\ P_{2,n+1} = s_1 P_{1,n} \\ \vdots \qquad\qquad \vdots \\ P_{K,n+1} = s_{K-1} P_{K-1,n} \,. \end{cases}$$

All the numbers f_k are nonnegative, while s_k satisfies $0 < s_k < 1$. At the end of the nineteenth and beginning of the twentieth century, mathematicians had taken the habit to write such systems of equations in the abbreviated form[1]

$$P_{n+1} = M P_n \,, \tag{21.1}$$

where P_n is the column vector $(P_{0,n}, \ldots, P_{K,n})$ and M is the square matrix (i.e., the table of numbers with $K+1$ rows and $K+1$ columns)

$$M = \begin{pmatrix} f_0 & f_1 & f_2 & \cdots & f_K \\ s_0 & 0 & 0 & \cdots & 0 \\ 0 & s_1 & 0 & \cdots & 0 \\ \vdots & \ddots & \ddots & \ddots & \vdots \\ 0 & \cdots & 0 & s_{K-1} & 0 \end{pmatrix}.$$

To understand the behaviour of system (21.1) as a function of time, Leslie looked for a geometrically increasing or decreasing solution $P_n = r^n V$. The number r and the vector V must satisfy

$$M V = r V \,. \tag{21.2}$$

In this case, r is called an "eigenvalue" and V an "eigenvector" of the matrix M. In other words, the problem is to find the age distribution V which at each time step is multiplied by a constant r. Following Lotka's terminology, such distributions are called "stable". Returning to more usual notations, equation (21.2) can be rewritten as

$$\begin{cases} f_0 V_0 + f_1 V_1 + \cdots + f_K V_K = r V_0, \\ s_0 V_0 = r V_1, \quad s_1 V_1 = r V_2, \quad \ldots, \quad s_{K-1} V_{K-1} = r V_K \,. \end{cases}$$

If follows from the last K equations that

---

[1] Meaning that $P_{k,n+1} = M_{k,0} P_{0,n} + M_{k,1} P_{1,n} + \cdots + M_{k,K} P_{K,n}$ for all k.

$$V_1 = \frac{s_0 V_0}{r}, \quad V_2 = \frac{s_0 s_1 V_0}{r^2}, \quad \ldots \ V_K = \frac{s_0 s_1 \cdots s_{K-1} V_0}{r^K}.$$

Replacing this in the first equation, simplifying by V_0 and multiplying by r^K, Leslie obtained the "characteristic equation"

$$r^{K+1} = f_0 r^K + s_0 f_1 r^{K-1} + s_0 s_1 f_2 r^{K-2} + \cdots + s_0 s_1 \cdots s_{K-1} f_K. \tag{21.3}$$

This is a polynomial equation in r of degree $K+1$. So there are $K+1$ real or complex roots $r_1, \ldots, r_{K+1}$. Moreover Leslie noticed (using Descartes' sign rule for polynomials) that there is just one real positive root. Call it r_1.

Leslie suggested also that, under most biologically realistic conditions (which can be made precise using the theory of Perron and Frobenius for nonnegative matrices), the eigenvalue r_1 is strictly bigger than the modulus of all the other real or complex eigenvalues (call them $r_2, \ldots, r_{K+1}$). Besides, all the roots of (21.3) are usually different. For each eigenvalue r_i, one can find an associated eigenvector. Let Q be the square matrix of size $K+1$ whose $K+1$ columns contain the eigenvectors respectively associated with $r_1, \ldots, r_{K+1}$, then $MQ = QD$, where D is the diagonal matrix $[r_1, \ldots, r_{K+1}]$. So $M = QDQ^{-1}$ and

$$P_n = M^n P_0 = QD^n Q^{-1} P_0.$$

Notice that D^n is the diagonal matrix $[(r_1)^n, \ldots, (r_{K+1})^n]$ and that

$$D^n/r_1^n \longrightarrow \mathscr{D} = [1, 0, \ldots, 0]$$

when $n \to +\infty$ because $r_1 > |r_i|$ for $i \neq 1$. Therefore, $P_n/(r_1)^n$ converges towards $Q \mathscr{D} Q^{-1} P_0$.

Each component of the age-structure vector P_n increases or decreases like $(r_1)^n$. If $r_1 > 1$, then the population increases exponentially. If $r_1 < 1$, then it decreases exponentially. From equation (21.3), one can easily show that the condition $r_1 > 1$ is true if and only if the parameter $\mathscr{R}_0$, defined by

$$\mathscr{R}_0 = f_0 + s_0 f_1 + s_0 s_1 f_2 + \cdots + s_0 s_1 \cdots s_{K-1} f_K,$$

is strictly bigger than 1. Notice that $s_0 s_1 \cdots s_{k-1}$ is the probability of surviving until at least age k. So the parameter $\mathscr{R}_0$ is the mean number of daughters born from one female throughout her life and is analogous to formulas (10.2), (12.2) and (16.9). The present model is a kind of discrete-time analogue of Lotka's work (see Chapter 10) and a generalization including age-dependent fertilities of Euler's work (see Chapter 3).

Leslie illustrated his method using data published by an American colleague on the fertility and survival coefficients f_k and s_k for the brown rat. After a few statistical operations to complete the data in a reasonable way, he obtained $\mathscr{R}_0 \simeq 26$.

Leslie's matrix formulation of problems in population dynamics is now used by many biologists. The computations are greatly simplified by modern computers and scientific software that can compute eigenvalues and eigenvectors of any matrix. One can easily compute both the parameter $\mathscr{R}_0$ and the growth rate r_1.

After the Second World War, Leslie used his method to compute the growth rate of other animal species: birds, beetles etc. He also worked on stochastic models, on models of competition between species and on the analysis of capture-recapture data. He retired in 1967. That same year, Charles Elton having also retired, the Bureau of Animal Population ceased to exist as an independent research centre and became part of the Department of Zoology at the University of Oxford. Leslie died in 1972.

Further reading

1. Anonymous: Dr P. H. Leslie. *Nature* **239**, 477–478 (1972)
2. Crowcroft, P.: *Elton's Ecologists, A History of the Bureau of Animal Population.* University of Chicago Press (1991). books.google.com
3. Leslie, P.H.: On the use of matrices in certain population mathematics. *Biometrika* **33**, 213–245 (1945)

Chapter 22
Percolation and epidemics (1957)

John Michael Hammersley was born in 1920 in Scotland, where his father worked for an American company exporting steel. He started studying at Emmanuel College of Cambridge University, but had to join the army in 1940. He worked at the improvement of computations for artillery. After finishing his studies in 1948, he became assistant at Oxford University in the group working on the design and analysis of experiments. In 1955 he joined the Atomic Energy Research Establishment in Harwell near Oxford.

Fig. 22.1 Hammersley (1920–2004)

Simon Ralph Broadbent was born in 1928. He studied engineering in Cambridge, mathematics at Magdalen College in Oxford (where he also wrote poetry) and started a PhD in statistics at Imperial College in London on "Some tests of departure from uniform dispersion". During his PhD he got some support from the British Coal Utilisation Research Association to investigate statistical problems that could be related to coal production.

In 1954 a symposium on Monte Carlo methods sponsored by the Atomic Energy Research Establishment was held at the Royal Statistical Society in London. These methods, initiated during the 1940s by John von Neumann, Stanisław Ulam and

Nicholas Metropolis at Los Alamos Laboratory, use stochastic computer simulations in order to estimate unknown mathematical quantities. Hammersley presented at the London symposium a paper that he had prepared in collaboration with Morton, a colleague from Harwell. The paper was also published in the *Journal of the Royal Statistical Society*. During the discussion following the presentation at the symposium, Broadbent mentioned an interesting problem that might be studied using some Monte Carlo method: given a regular network of pores in two or three dimensions such that two neighbouring pores are connected with a probability p, what proportion of the network would be filled by a gas if it were introduced through one of these pores? Broadbent was in fact thinking at the design of gas masks for coal miners and in particular about the size of the pores that was necessary for their functioning.

Hammersley then started to work with Broadbent on this gas mask problem. They realized that it was just a prototype of a family of problems that had not yet been studied: the deterministic propagation of a "fluid" (the meaning depending on the context) in a random medium. Hammersley called it "percolation", by analogy with what happens in a coffee pot. At the Atomic Energy Research Establishment, Hammersley also had access to some of the most powerful computers of his time to test Monte Carlo methods on percolation problems.

In 1957 Broadbent and Hammersley finally published the first article on the mathematical theory of percolation. Among the examples they considered, one was a model of population dynamics, namely the propagation of an epidemic in an orchard. The trees of a very big orchard are assumed to be placed at the nodes of a square network. Each of the four closest trees of a given infected tree has a probability p to be also infected. The question is whether a large number of trees will be infected or if the epidemic will stay localized. This depends of course on the probability p, which in turn is linked to the distance separating the trees, i.e. the width of the network mesh.

Broadbent and Hammersley looked at the limiting case where the orchard is infinite and covers the entire plane, with just one infected tree at the beginning. Let $f(p)$ be the probability that an infinite number of trees become infected from this source. One expects $f(p)$ to be an increasing function of p with $f(0) = 0$ and $f(1) = 1$. Their main result was that there is a critical probability p^*, $0 < p^* < 1$, such that:

- if $p < p^*$, then $f(p) = 0$ so only a finite number of trees are infected;
- if $p > p^*$, then $f(p) > 0$ and an infinite number of trees may be infected.

The proof involves a comparison with the number of different "self-avoiding walks" in the plane starting from the source of infection. These walks go through a certain number of neighbouring trees (recall that each tree has four neighbours) without visiting any tree more than once. An n-stepped self-avoiding walk is a path of infection with a probability p^n since the infection can be transmitted from each visited tree to the next with a probability p. Now let $q(j,n)$ be the probability that, among all n-stepped self-avoiding walks, there are exactly j such walks that are paths of infection. If there is an infinite number of infected trees, then for all integer n there exists at least one n-stepped self-avoiding walk that is a path of infection. So

$$0 \le f(p) \le \sum_{j=1}^{\infty} q(j,n) \le \sum_{j=1}^{\infty} j\, q(j,n)$$

for all n. But $\sum_{j=1}^{\infty} j\, q(j,n)$ is the expected number of n-stepped self-avoiding walks that are paths of infection. This number is equal to $p^n s(n)$, where $s(n)$ is the total number of n-stepped self-avoiding walks. Hammersley could show in a companion paper that $s(n)$ grows like $e^{\kappa n}$ as $n \to +\infty$, where κ is called the connective constant. If $p < e^{-\kappa}$, then $p^n s(n)$ tends to 0 as $n \to +\infty$ and $f(p) = 0$. Thus $p^* \ge e^{-\kappa} > 0$.

In practice it is therefore better if the trees are not too close to keep p below p^* in case of an epidemic. But the closer the trees, the higher the production per hectare. A compromise has to be found.

As Broadbent and Hammersley noticed, there is a certain similarity between the existence of a critical probability in percolation processes and the existence of a threshold in branching processes (see Chapter 7).

One can try to estimate numerically the critical probability p^*. For this purpose, fix a value for p and approximate the infinite network by a finite square network of size $N \times N$ with N sufficiently large. Assume for example that the tree in the middle of the network is infected. With a computer, one can choose randomly which trees can infect other trees. Fig. 22.2 and Fig. 22.3 show the randomly chosen paths of infection using edges as in a graph. In Fig. 22.2, p is smaller than p^*. In Fig. 22.3, p is bigger than p^*. One can easily determine which trees can be infected, namely those that can be reached by a path of edges starting from the infected tree in the centre. They are marked by small black squares in the figures.

One can then check if the epidemic has reached at least the border of the $N \times N$ network. If this is so and if N is large enough, one can consider that the number of infected trees is "almost infinite". Repeating this kind of simulation many times, one can find an approximate value of the probability $f(p)$ that the number of infected trees is infinite (this is the Monte Carlo method). Finally, letting p vary between 0 and 1, one can get an approximation of the threshold p^*, which is the smallest value such that $f(p) > 0$ if $p > p^*$.

The article by Broadbent and Hammersley contained only the proof of the existence of the threshold p^*. During the following years Hammersley continued to develop the mathematical theory of percolation, while Broadbent turned to other subjects. With the development of computers in the 1970s, it became easier to run the simulations described above (Fig. 22.4). It was then conjectured that $p^* = 1/2$. This result was finally proved in 1980 by Harry Kesten from Cornell University.

Fig. 22.2 Percolation with $p = 0.4$.

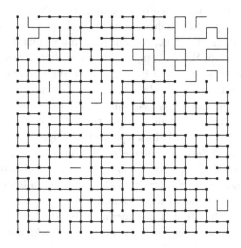

Fig. 22.3 Percolation with $p = 0.55$.

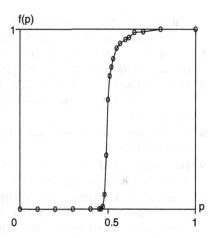

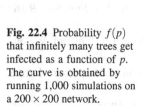

Fig. 22.4 Probability $f(p)$ that infinitely many trees get infected as a function of p. The curve is obtained by running 1,000 simulations on a 200×200 network.

Between 1959 and 1969 Hammersley worked for the Institute of Economics and Statistics at Oxford University. He became a fellow of Trinity College. In 1964 he published in collaboration with David Handscomb a book entitled *Monte Carlo Methods*. He was elected to the Royal Society in 1976. He retired in 1987 but continued to visit the Oxford Centre for Industrial and Applied Mathematics. He died in 2004.

Broadbent got his PhD at Imperial College in 1957. He found a job in an industrial company, the United Glass Bottle Manufacturers. After ten years in industry he began to work in a news agency, the London Press Exchange, which did scientific readership studies. The agency was bought in 1969 by Leo Burnett, an American advertising company. Broadbent worked on how to measure the effectiveness of advertising and published several books on that subject: *Spending Advertising Money* (1975), *Advertising Budget* (1989), *Accountable Advertising* (1997) and *When to Advertise* (1999). In 1980 he helped start the Advertising Effectiveness Awards. He spent several years at Leo Burnett's head office in Chicago as director of brand economics. He also ran his own consultancy, BrandCon Limited. He died in 2002.

Further reading

1. Grimmett, G., Welsh, D.: John Michael Hammersley. *Biogr. Mem. Fellows R. Soc.* **53**, 163–183 (2007)
2. Broadbent, S.R.: Discussion on symposium on Monte Carlo methods. *J. R. Stat. Soc. B* **16**, 68 (1954)
3. Broadbent, S.R., Hammersley, J.M.: Percolation processes I: Crystals and mazes. *Proc. Camb. Philos. Soc.* **53**, 629–641 (1957)
4. Broadbent, T.: Simon Broadbent – The man with a sense of fun who gave advertising a value. *Campaign*, 26 April 2002. campaignlive.co.uk

5. Hammersley, J.M.: Percolation processes II: The connective constant. *Proc. Camb. Philos. Soc.* **53**, 642–645 (1957)

6. Hammersley, J.M.: Percolation processes: lower bounds for the critical probability. *Ann. Math. Stat.* **28**, 790–795 (1957)

7. Hammersley, J.M.: Origins of percolation theory. In: Deutscher, G., Zallen, R., Adler, J. (eds.) *Percolation Structures and Processes*, pp. 47–57. Israel Physical Society (1983)

8. Hammersley, J.M., Morton, K.W.: Poor man's Monte Carlo. *J. R. Stat. Soc. B* **16**, 23–38 (1954)

9. Hammersley, J.M., Handscomb, D.C.: *Monte Carlo Methods*. Fletcher & Son, Norwich (1964). books.google.com

10. Kesten, H.: The critical probability of bond percolation on the square lattice equals 1/2. *Comm. Math. Phys.* **74**, 41–59 (1980)

11. Metropolis, N., Ulam, S.: The Monte Carlo method. *J. Amer. Stat. Assoc.* **44**, 335–341 (1949)

Chapter 23
Game theory and evolution (1973)

John Maynard Smith was born in London in 1920. His father, who was a surgeon, died when he was eight. Maynard Smith studied at Eton College and turned to engineering studies at Trinity College, Cambridge University. He was then a member of the Communist Party of Great Britain. In 1939, when war broke out, he tried to volunteer for the army but was rejected because of his poor eyesight. He finished his engineering studies and worked for some years on the design of military aircraft. Finally he decided to turn to biology, studying genetics at University College in London with Haldane as a supervisor. He became a lecturer in zoology in 1952. He left the Communist Party after the 1956 events in Hungary. His first book, entitled *The Theory of Evolution*, was published in 1958. In 1965 he became professor of biology at the newly founded University of Sussex. He then published two other books: *Mathematical Ideas in Biology* (1968) and *On Evolution* (1972).

Fig. 23.1 Maynard Smith
(1920–2004)

George R. Price was born in 1922 in the USA. He studied chemistry at the University of Chicago, getting a Ph.D. in 1946 after having worked on the Manhattan Project, building the atomic bomb. In 1950 he became associate researcher in medicine at the University of Minnesota. He later worked as an independent journalist for various magazines before returning to research at IBM. In 1967, after having been treated for thyroid cancer, he settled in England and turned to the study of a completely different subject: evolutionary biology. He worked in London at the Galton Laboratory of University College from 1968. His first paper in this new area, "Selection and covariance", was published with the help of W. D. Hamilton in a 1970 issue of *Nature* and contained what is now called Price's equation.

Price also submitted another paper to *Nature*, this time on animal conflicts. But it did not have the right format for this journal. So Maynard Smith, who was the reviewer, suggested preparing a shorter version. Price started to work on something else while Maynard Smith began to develop Price's idea on his own. Finally Maynard Smith and Price published a joint article entitled "The logic of animal conflict", which *Nature* published in 1973. The article made an interesting contribution to the use of game theory in evolutionary biology. Before that, game theory had been mainly developed for economics and politics, especially after the 1944 book by John von Neumann and Oskar Morgenstern entitled *Theory of Games and Economic Behavior*. The starting point of Maynard Smith and Price was the following question: how is it that in conflicts between animals of the same species, the "weapons" at their disposal (horns, claws, venom etc.) are rarely used to kill? Following Darwin's ideas on the struggle for life, more aggressive animals should win more combats and have a larger number of offspring, leading to an escalation in the use of "weapons". Notice that this was the time of the Cold War so the subject also had a political flavour.

Maynard Smith and Price imagined a sequence of games in which two animals can enter in competition for a resource, for example a territory in a favourable habitat. In the simplified presentation that Maynard Smith would use in his 1982 book *Evolution and the Theory of Games*, each animal adopts either the "hawk strategy" or the "dove strategy". In what follows we talk simply about hawks and doves, but we mean strategies adopted by animals of the same species. Let $V > 0$ be the value of the resource, meaning that if $\mathscr{R}_0$ is the normal average number of offspring of an animal, the winner of the competition has on average $\mathscr{R}_0 + V$ offspring.

If a hawk meets another hawk, they fight for the resource: the winner gets the resource of value V, the loser suffers a "cost" $C > 0$. Each of the two hawks has a probability equal to $1/2$ of winning the competition and the same probability of losing. The expected payoff from a fight between two hawks is therefore $\frac{1}{2}(V - C)$ for the two competitors. If, however, a hawk meets a dove, then the hawk gets the resource V, the dove escapes without fighting and the cost is 0. Finally, if two doves meet, one of them gets the resource V, the other escapes without fighting and at no cost. Each of the two doves having the same probability $1/2$ of winning, the expected payoff when two doves meet is therefore $V/2$. The payoffs can be summarized as in Table 23.1.

More generally, one can imagine fights between individuals that can adopt one of two strategies, call them 1 and 2, with a matrix of expected payoffs $(G_{i,j})_{1 \le i,j \le 2}$. In the example above, hawks follow strategy 1, doves follow strategy 2, $G_{1,1} = \frac{1}{2}(V - C)$, $G_{1,2} = V$, $G_{2,1} = 0$ and $G_{2,2} = V/2$. In the original article of 1973, Maynard Smith and Price had in fact already used computer simulations to test more than two

Table 23.1 Expected payoffs of the hawk–dove game.

	A hawk	A dove
payoff of a hawk against...	$\frac{1}{2}(V-C)$	V
payoff of a dove against...	0	$V/2$

possible strategies (these were called "hawk", "mouse", "bully", "retaliator" and "prober-retaliator").

Imagine now a large population of animals of the same species with a proportion x_n of hawks and a proportion $1 - x_n$ of doves in generation n. Hawks in generation n have an average number of offspring equal to

$$R_1(n) = \mathcal{R}_0 + x_n G_{1,1} + (1 - x_n) G_{1,2}. \qquad (23.1)$$

Similarly, doves have an average number of offspring equal to

$$R_2(n) = \mathcal{R}_0 + x_n G_{2,1} + (1 - x_n) G_{2,2}. \qquad (23.2)$$

The average number of offspring in the entire population is therefore

$$R(n) = x_n R_1(n) + (1 - x_n) R_2(n).$$

Forgetting about the possible subtleties due to sexual reproduction, we see that the proportion of hawks in the next generation is

$$x_{n+1} = x_n R_1(n)/R(n). \qquad (23.3)$$

Hence, $x_{n+1} > x_n$ if $R_1(n) > R(n)$ and $x_{n+1} < x_n$ if $R_1(n) < R(n)$. There are three possible steady states: $x = 0$, $x = 1$ and

$$x^* = \frac{G_{1,2} - G_{2,2}}{G_{2,1} - G_{1,1} + G_{1,2} - G_{2,2}}$$

provided $0 < x^* < 1$. In the hawk–dove game, $x^* = V/C < 1$ provided $V < C$.

Indeed, $x = 0$ is an obvious steady state of (23.3). If $x \neq 0$ is another steady state, then $R_1 = R = xR_1 + (1-x)R_2$. So either $x = 1$ or $R_1 = R_2$. The latter possibility is equivalent to $xG_{1,1} + (1-x)G_{1,2} = xG_{2,1} + (1-x)G_{2,2}$, which gives the steady state x^*.

The steady state $x = 1$ corresponds to a population with 100% of individuals following strategy 1. This steady state is stable if it cannot be invaded by a few individuals following strategy 2. From (23.3), we see that this condition is equivalent to having $R_1(n) > R(n)$ for all x_n sufficiently close to 1. Since $R(n) = x_n R_1(n) + (1 -$

x_n) $R_2(n)$, the condition becomes $R_1(n) > R_2(n)$ for all x_n sufficiently close to 1. Looking at the expressions (23.1)–(23.2) of R_1 and R_2, we arrive at the conclusion that $x = 1$ is stable if and only if one of the two following conditions is satisfied:

- $G_{1,1} > G_{2,1}$;
- $G_{1,1} = G_{2,1}$ and $G_{1,2} > G_{2,2}$.

If so, strategy 1 is said to be an *evolutionarily stable strategy* (ESS in short). In the hawk–dove game, the condition $G_{1,2} > G_{2,2}$ is always true. So the hawk strategy is evolutionarily stable if and only if $G_{1,1} \geq G_{2,1}$, i.e. $V \geq C$.

The steady state $x = 0$ corresponds to a population with all individuals following strategy 2. This situation is symmetric to the previous one if we exchange indices 1 and 2. In the hawk–dove game, we have $G_{1,2} = V > G_{2,2} = V/2$ so the steady state $x = 0$ is always unstable. Introducing a small number of hawks in a population of doves would lead to a progressive invasion by the hawks.

Similarly, one can show that the third steady state x^*, provided $0 < x^* < 1$, is always stable. In the hawk–dove game, $x^* = V/C$ corresponds to a mixed population with both hawks and doves.

In conclusion, there are two cases in the hawk–dove game. If $V \geq C$, i.e. if the value of the resource is bigger than the possible cost, then the population tends to a steady state with hawks but no doves, whatever the initial condition $x(0)$ with $0 < x(0) < 1$. The hawk strategy is then an evolutionarily stable strategy. If, in contrast, $V < C$, then the population tends to a mixed steady state with a proportion x^* of hawks and a proportion $1 - x^*$ of doves. So the model does give an explanation of why individuals with less aggressive behaviours can survive when $V < C$. The formula $x^* = V/C$ shows moreover that the higher the cost C for losers, the smaller the proportion x^* of hawks in the population. Hence species with the most dangerous "weapons" seldom use them for intraspecific fights: they prefer inoffensive ritual fights, where competing animals try to impress each other but avoid real fights that could cause injuries.

The original 1973 article by Maynard Smith and Price discussed the concept of evolutionarily stable strategy and used mainly computer simulations of animal contests, recording the payoffs of different strategies. The approach using dynamical equations such as (23.3) was developed somewhat later, in particular by Taylor and Jonker. Since then many authors have applied ideas from game theory to questions in evolutionary biology or conversely have applied dynamical evolutionary approaches to more classical problems in game theory. Besides questions concerning animal conflicts, one can cite for example problems of parental investment or of sex ratio (the ratio between the number of males and females at birth), the latter having been studied already by Carl Düsing in 1884 and by Ronald Fisher in his 1930 book on *The Genetical Theory of Natural Selection*. Some other models focus on the dynamic aspects of the "prisoner's dilemma" or of the "rock-paper-scissors" game. It was also realized that the concept of evolutionarily stable strategy is closely related to the concept of Nash equilibrium in game theory.

Price, who had been a convinced atheist, had a mystical experience in 1970 and converted to the Christian faith. He gave up his research in 1974 because he felt that "the sort of theoretical mathematical genetics [he] was doing wasn't very relevant to human problems". He gave all his belongings to homeless people and committed suicide a few months later.

Maynard Smith, in contrast, continued this line of thought and was elected to the Royal Society in 1977. He published many books: *Models in Ecology* (1974), *The Evolution of Sex* (1978), *Evolution and the Theory of Games* (1982), *The Problems of Biology* (1986), *Did Darwin Get it Right?* (1988) and *Evolutionary Genetics* (1989). He also published in collaboration with E. Szathmáry *The Major Transitions in Evolution* (1995) and *The Origins of Life: From the Birth of Life to the Origin of Language* (1999). He retired in 1985. In 1999 he received the Crafoord prize in biosciences from the Royal Swedish Academy of Sciences for his "fundamental contributions to the conceptual development of evolutionary biology". In 2003 he published in collaboration with D. Harper *Animal Signals*. He died in Sussex in 2004.

Further reading

1. Charlesworth, B., Harvey, P.: John Maynard Smith, 6 January 1920–19 April 2004. *Biog. Mem. Fellows R. Soc.* **51**, 253–265 (2005)
2. Edwards, A.W.F.: Carl Düsing (1884) on the regulation of the sex-ratio. *Theor. Pop. Biol.* **58**, 255–257 (2000)
3. Frank, S.A.: George Price's contributions to evolutionary genetics. *J. Theor. Biol.* **175**, 373–388 (1995)
4. Maynard Smith, J., Price, G.R.: The logic of animal conflict. *Nature* **246**, 15–18 (1973)
5. Maynard Smith, J.: *Evolution and the Theory of Games*. Cambridge University Press (1982). books.google.com
6. Schwartz, J.: Death of an altruist: Was the man who found the selfless gene too good for this world? *Lingua Franca* **10**, 51–61 (2000). bio.kuleuven.be
7. Sigmund, K.: John Maynard Smith and evolutionary game theory. *Theor. Pop. Biol.* **68**, 7–10 (2005)
8. Taylor, P.D., Jonker, L.B.: Evolutionary stable strategies and game dynamics. *Math. Biosci.* **40**, 145–156 (1978)
9. Von Neumann, J., Morgenstern, O.: *Theory of Games and Economic Behavior*. Princeton University Press (1944). www.archive.org

Chapter 24
Chaotic populations (1974)

Robert McCredie May was born in 1936 in Australia. After studying theoretical physics and receiving a PhD from the University of Sydney in 1959, he spent two years in the department of applied mathematics at Harvard University. Back in Australia, he became professor of theoretical physics. In 1971, while visiting the Institute for Advanced Study in Princeton, he changed his research subject and started to focus on animal population dynamics. In 1973 he became professor of zoology in Princeton. The same year he published a book entitled *Stability and Complexity in Model Ecosystems*.

Fig. 24.1 Robert M. May

In 1974 May published in *Science* an article entitled "Biological populations with nonoverlapping generations: stable points, stable cycles and chaos", in which he showed that very simple mathematical models in population dynamics can behave in a chaotic way.

To understand the origin of this problem, one has to go back about ten years in time. In 1963 Edward Lorenz, an American meteorologist working at the Massachusetts Institute of Technology (M.I.T.), had noticed while making numerical

N. Bacaër, *A Short History of Mathematical Population Dynamics*,
DOI 10.1007/978-0-85729-115-8_24, © Springer-Verlag London Limited 2011

simulations on his computer that a simplified model of the atmosphere with only three differential equations could behave in a very surprising way: a tiny change of the initial conditions could change completely the final result of a simulation and therefore also meteorological forecasts. The mathematician Henri Poincaré, after having worked on the motion of planets in the Solar System, had in fact already thought about this possibility at the beginning of the twentieth century, long before the computer age. But at the beginning of the 1970s, only a few researchers were starting to look at this strange property more closely. At the University of Maryland, James Yorke was thinking about the work of Lorenz and introduced the term "chaos" in this context. The article[1] he wrote with his student Tien-Yien Li, entitled *Period three implies chaos*, appeared in 1975.

On his side, May was focusing on the model

$$p_{n+1} = p_n + a p_n (1 - p_n/K), \tag{24.1}$$

where a and K are positive parameters and p_n stands for the size of an animal population in year n. When p_n is small compared to the carrying capacity K, the dynamics is close to a geometric growth $p_{n+1} \simeq (1+a) p_n$. The full equation is a kind of discrete-time analogue of the logistic equation introduced by Verhulst (see Chapter 6). But unlike the latter, May showed that the discrete-time equation can have a much more surprising behaviour, which is easy to observe with a simple pocket calculator doing additions and multiplications (Fig. 24.2). Maynard Smith had already considered equation (24.1) in his 1968 book *Mathematical Ideas in Biology*. But despite having tried a few numerical values for a, he had not realized that there was something special.

Figure 24.2, which is similar to the one in May's 1974 article, shows that the population p_n converges to a steady state when $0 < a < 2$. When $2 < a \leq 2.449$ (the upper bound 2.449 is an approximation), the population p_n tends to a cycle of period 2. When $2.450 \leq a \leq 2.544$, the population p_n tends to a cycle of period 4. When $2.545 \leq a \leq 2.564$, p_n tends to a cycle of period 8, etc. The intervals of the parameter a for which p_n tends to a cycle of period 2^n get smaller as n increases and never exceed 2.570. When $a \geq 2.570$, p_n can behave in a "chaotic" way.

In 1976 May wrote a review of the problem, published in *Nature*, entitled *Simple mathematical models with very complicated dynamics*. There he collected not only his own results but also those of other researchers. First, setting $x_n = \frac{a p_n}{K(1+a)}$ and $r = 1+a$ (so that $r > 1$), we see that equation (24.1) can be rewritten in the more simple form

$$x_{n+1} = r x_n (1 - x_n) . \tag{24.2}$$

For this equation to have a meaning in population dynamics, x_n should be nonnegative for all n. So we assume that the initial condition x_0 satisfies $0 \leq x_0 \leq 1$ and that $r \leq 4$. The latter condition ensures that the right-hand side of (24.2) stays between 0 and 1. Remarkably the chaotic case $r = 4$ had already been used as a random num-

---

[1] Remarkably a more general result was proved by O. M. Sharkovsky in 1964, but his article published in a Ukrainian mathematics journal was not well known.

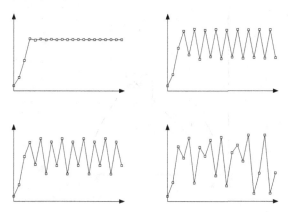

Fig. 24.2 In all the figures: n is on the horizontal axis, p_n on the vertical axis and $p_0 = K/10$. The lines are obtained by joining the points with coordinates (n, p_n). Top left: $0 < a < 2$ (steady state). Top right: $2 < a \leq 2.449$ (period 2 cycle). Bottom left: $2.450 \leq a \leq 2.544$ (period 4 cycle). Bottom right: $2.570 \leq a \leq 3$ (possibly chaos).

ber generator by Stanisław Ulam and John von Neumann as early as 1947. If we introduce the function

$$f(x) = rx(1-x),$$

then equation (24.2) can be rewritten as $x_{n+1} = f(x_n)$ and the steady states are the solutions of $x = f(x)$. Graphically these are the intersections of the curves $y = f(x)$ and $y = x$ (Fig. 24.3). Notice that $x = 0$ is always a steady state. Since $r > 1$, there is also another steady state $x^* > 0$ such that $x^* = rx^*(1 - x^*)$, i.e., $x^* = 1 - 1/r$.

Because $r > 1$, the steady state $x = 0$ is unstable. Indeed, when x_n is close to 0, we have $x_{n+1} \simeq rx_n$. So x_n tends to move away from 0. As for the steady state x^*, it is locally stable only for $1 < r < 3$.

Indeed, set $y_n = x_n - x^*$. Then (24.2) is equivalent to $y_{n+1} = (2 - r - ry_n)y_n$. If x_n is close to x^*, then y_n is close to 0 and $y_{n+1} \simeq (2 - r)y_n$. But if $y_{n+1} = ky_n$, then $y_n = k^n y_0$ so that $y_n \to 0$ when $n \to \infty$ if and only if $-1 < k < 1$. Here the steady state x^* is locally stable if and only if $-1 < 2 - r < 1$, i.e. $1 < r < 3$.

When $1 < r < 3$, one can show that for all initial conditions $0 < x_0 < 1$, the sequence x_n really tends to x^* (Fig. 24.3a). But what happens when $3 < r \leq 4$? To answer this question, notice first that $x_{n+2} = f(x_{n+1}) = f(f(x_n))$. Introduce the function

$$f_2(x) = f(f(x)) = r^2 x(1-x)\left(1 - rx(1-x)\right)$$

and consider the solutions of the equation $x = f_2(x)$, which are called fixed points of the function $f_2(x)$. Graphically these are the intersections of the curves $y = f_2(x)$

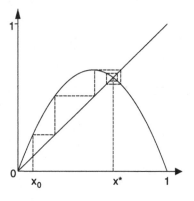

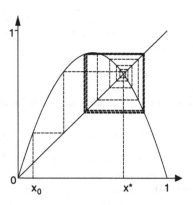

Fig. 24.3 The function $y = f(x)$, the straight line $y = x$, the steady state x^* and the sequence defined by $x_{n+1} = f(x_n)$. Top: $r = 2.75$, the sequence tends to x^*. Bottom: $r = 3.4$, the steady state x^* is unstable and the sequence tends to a cycle of period 2.

and $y = x$ (Fig. 24.4). If $x = f(x)$, then $x = f(f(x)) = f_2(x)$. So $x = 0$ and $x = x^*$ are also fixed points of the function $f_2(x)$. But when $r > 3$, the function $f_2(x)$ has two other fixed points, x_- and x_+, such that $f(x_-) = x_+$ and $f(x_+) = x_-$.

Indeed we notice that $f_2'(x) = f'(f(x)) f'(x)$ so that $f_2'(x^*) = [f'(x^*)]^2$. But $f'(x) = r(1 - 2x)$ and $x^* = 1 - 1/r$. So $f'(x^*) = 2 - r$ and $f_2'(x^*) = (2 - r)^2$. Hence the slope of the function $f_2(x)$ at $x = x^*$ is such that $f_2'(x^*) > 1$ if $r > 3$. But since $f_2(0) = 0$, $f_2'(0) = r^2 > 1$ and $f_2(1) = 0$, we see in Fig. 24.4b that there are necessarily two other solutions x_- and x_+ of equation $x = f_2(x)$, with

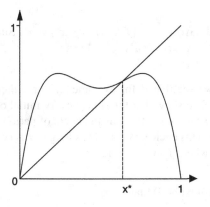

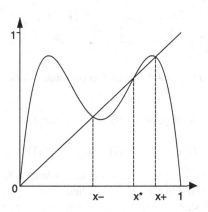

Fig. 24.4 The curves $y = f_2(x) = f(f(x))$ and $y = x$ and the steady state x^*. Top: $r = 2.75$. Bottom: $r = 3.4$ and the two other solutions x_- and x_+ of equation $x = f_2(x)$.

$0 < x_- < x^*$ and $x^* < x_+ < 1$. Another way of arriving at the same conclusion consists in solving equation $x = f_2(x)$, which is a polynomial equation of degree 4 with two known roots: $x = 0$ and $x = x^*$. The two other solutions x_- and x_+ are the roots of the polynomial

$$x^2 - \frac{1+r}{r} x + \frac{1+r}{r^2} = 0. \tag{24.3}$$

They are real if the discriminant is positive, i.e. if $r > 3$. Since $f_2(f(x_-)) = f(f(f(x_-))) = f(f_2(x_-)) = f(x_-)$, the point $f(x_-)$ is also a fixed point of $f_2(x)$. But $f(x_-) \neq x_-$ because x_- is not a fixed point of $f(x)$. And $f(x_-) \neq x^*$, otherwise we would have $x_- = f(f(x_-)) = f(x^*) = x^*$. Since $f(x_-) \neq 0$, we conclude that $f(x_-) = x_+$. Similarly, $f(x_+) = x_-$.

Hence for $r > 3$, we see that if for example $x_0 = x_-$, then $x_1 = x_+$, $x_2 = x_-$, $x_3 = x_+$, etc. One can also show that for almost every initial condition $0 < x_0 < 1$, the sequence x_n tends as $n \rightarrow +\infty$ towards the cycle of period 2 x_-,x_+,x_-,x_+, etc. (Fig. 24.3b and 24.4b). This cycle stays stable as long as r is below the critical value $r_1 = 1 + \sqrt{6} \simeq 3.449$, where $f_2'(x_-) = -1$.

Indeed, we see by using (24.3) that

$$f_2'(x_-) = f'(f(x_-))f'(x_-) = f'(x_+)f'(x_-)$$
$$= r^2(1-2x_+)(1-2x_-) = r^2(1-2(x_+ + x_-)+4x_+x_-)$$
$$= r^2\left(1 - 2\frac{1+r}{r} + 4\frac{1+r}{r^2}\right) = -r^2 + 2r + 4.$$

So $f_2'(x_-) = -1$ if $-r^2 + 2r + 5 = 0$ and in particular if $r = 1 + \sqrt{6}$.

For $r_1 < r < r_2$, a cycle of period 4 becomes stable: four new fixed points of the function

$$f_4(x) = f_2(f_2(x)) = f(f(f(f(x))))$$

appear (Fig. 24.5). For $r_2 < r < r_3$, it is a cycle of length 8, etc. The numbers r_n tend to a limit $r_\infty \simeq 3.570$ when $n \rightarrow +\infty$. When $r_\infty < r \leq 4$, the system can even be chaotic! Figure 24.6 shows the bifurcation diagram[2], which gives an idea of the dynamics' complexity.

R. M. May concluded by emphasizing that even very simple dynamical systems could have a very complicated behaviour. This is not specific to equation $x_{n+1} = rx_n(1-x_n)$. The same "period doubling cascade" leading to chaos appears for other equations with a function $f(x)$ having the shape of a "bump". This is the case for example with another equation used in population biology: $x_{n+1} = x_n \exp(r(1 - x_n))$.

This study suggests that one should not be surprised if many data sets concerning population dynamics are difficult to analyze. The model also shows that the distinction between deterministic and stochastic models is not as clear as previously thought: even with a simple deterministic model, it can be impossible to make long-term forecasts if the parameters are in the chaotic regime.

In 1979 May was elected to the Royal Society. From 1988 till 1995, he was professor at Oxford University and at Imperial College in London. From 1995 till

---

[2] This diagram was obtained by plotting for each given value of r the points with coordinates (r,x_{200}), $(r,x_{201}),\ldots,(r,x_{220})$, where $x_{n+1} = f(x_n)$ and $x_0 = 0.1$. If x_n tends to a steady state, we see only one point in the diagram. If x_n tends to a cycle of period 2, we see two points etc.

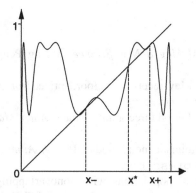

Fig. 24.5 The curve $y = f_4(x)$ when $r = 3.5$ and the line $y = x$. Beside x^*, x_+ and x_-, there are four other fixed points, which are not easy to distinguish.

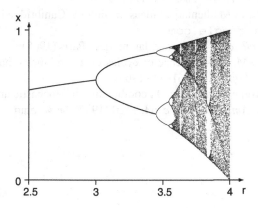

Fig. 24.6 Bifurcation diagram of equation (24.2).

2000, he was chief scientific adviser to the British government. In 1996 he received the Craford prize "for his pioneering ecological research concerning theoretical analysis of the dynamics of populations, communities and ecosystems". From ecology he turned to epidemiology and immunology, publishing two books: *Infectious Diseases of Humans* (1991, with Roy Anderson) and *Virus Dynamics, The Mathematical Foundations of Immunology and Virology* (2000, with Martin Nowak). The latter book analyzes the interaction between the cells of the immune system and HIV (the virus causing AIDS) as some kind of predator–prey system (see Chapter 13). From 2000 till 2005, May was president of the Royal Society. He was knighted in 1996 and became a life peer in 2001.

Further reading

1. Gleick, J.: *Chaos, Making a New Science*. Viking Penguin Inc., New York (1987)
2. Levin, S.A.: Robert May receives Crafoord prize. *Not. Amer. Math. Soc.* **43**, 977–978 (1996). www.ams.org
3. Li, T.Y., Yorke, J.A.: Period three implies chaos. *Amer. Math. Monthly* **82**, 985–992 (1975)
4. Lorenz, E.N.: Deterministic nonperiodic flow. *J. Atmosph. Sci.* **20**, 130–141 (1963). ams.allenpress.com
5. May, R.M.: Biological populations with nonoverlapping generations: stable points, stable cycles and chaos. *Science* **186**, 645–647 (1974)
6. May, R.M.: Simple mathematical models with very complicated dynamics. *Nature* **261**, 459–467 (1976)
7. May, R.M., Oster, G.F.: Bifurcations and dynamic complexity in simple ecological models. *Amer. Natur.* **110**, 573–599 (1976)
8. Maynard Smith, J.: Mathematical Ideas in Biology. Cambridge University Press (1968). books.google.com
9. Poincaré, H.: *Science et Méthode*. Flammarion, Paris (1908)
10. Sharkovsky, O.M.: Co-existence of cycles of a continuous mapping of a line onto itself. *Ukr. Math. J.* **16**, 61–71 (1964)
11. Ulam, S.M., von Neumann, J.: On combination of stochastic and deterministic processes. *Bull. Amer. Math. Soc.* **53**, 1120 (1947). www.ams.org

Chapter 25
China's one-child policy (1980)

Song Jian[1] was born in 1931 in Rongcheng in the Chinese province of Shandong. During the 1950s he studied in the Soviet Union at the Bauman Moscow State Technical University and at the Mathematics and Mechanics Department of Moscow State University. He then returned to China and became the head of the Office of Cybernetic Research in the Mathematics Institute of the Chinese Academy of Sciences. He was a specialist of the application of control theory to the guidance of missiles. He also worked for the Seventh Machine Building Ministry, which was later renamed Ministry of Aerospace. In 1978 he began to focus on the links between control theory and demography.

Fig. 25.1 Song Jian

To understand the context of Song Jian's work on population dynamics, one should first give an idea of what "control theory" is. It is the study of dynamical systems whose behaviour depends on some parameters that can be modified as time goes by in order to optimize a given criterion. This theory had been particularly developed in connection with space programs in the USA and in the USSR. Indeed, engineers had to "control" the trajectory of space shuttles in order to bring satellites to their orbit around the Earth. But applications were not limited to physical or en-

---

[1] Song is the family name. It is always written first in Chinese.

N. Bacaër, *A Short History of Mathematical Population Dynamics*,
DOI 10.1007/978-0-85729-115-8_25, © Springer-Verlag London Limited 2011

gineering problems. Birth control policies could also be considered as some kind of optimal control problem in the mathematical sense.

One should also mention the essay entitled *The Limits to Growth: A Report for the Club of Rome's Project on the Predicament of Mankind*, published in 1972 and written by a group from the Massachusetts Institute of Technology (M.I.T.). This study was based on a mathematical model of the world's economic growth that took into account natural resources, population size and pollution. The report suggested that the world's economy was heading towards a catastrophe by exhaustion of non-renewable resources, by lack of food for the population or by an excess of pollution. The voluntary limitation of births was one of the proposed solutions. In summary it was a kind of modern version of Malthus' theses. The report received a large echo in the West during the 1970s.

Since the founding of the People's Republic in 1949, the Chinese birth rate had been very high except during the catastrophic "Great Leap Forward". In the mid-1970s China was slowly recovering from the Cultural Revolution. Family planning urged women to delay births, to increase the time between two consecutive births and to have less children. Deng Xiaoping, who emerged as the new leader after Mao Zedong's death in 1976, started the policy of "Four Modernizations" in 1978: agriculture, industry, science and technology, and national defence. The size and the growth of the Chinese population were then perceived as important obstacles to these modernizations. Scientists that had been working until then on military applications were encouraged to find solutions for this difficult problem.

With this background, Song Jian went in 1978 to Helsinki for a congress of the International Federation of Automatic Control. He noticed there that some researchers in Europe had been trying to apply control theory to population problems with the idea that a strict birth control could eventually prevent the catastrophes announced by the report on *The Limits to Growth*. Back in China he set up a small team, including his colleague Yu Jingyuan and the computer expert Li Guangyuan, to apply this kind of mathematical modelling to data concerning the Chinese population. At that time scientific communication between China and the rest of the world was scarce. The team redeveloped the equations describing the evolution of a population's age structure, in the same way Lotka and McKendrick had done (see Chapters 10 and 16). Using a continuous-time model, call

- $P(x,t)$ the population aged x at time t;
- $m(x)$ the mortality at age x;
- $P_0(x)$ the population's age structure at time $t = 0$;
- $b(t)$ the total fertility of women at time t, i.e. the mean number of children a woman would have during her life if age-specific fertility remained what it is at time t;
- f the proportion of female births;
- $h(x)$ the probability distribution of the age of the mother when a child is born ($\int_0^\infty h(x)\,dx = 1$).

With these notations and hypotheses, the evolution of the age structure can be modelled by the partial differential equation

$$\frac{\partial P}{\partial t}(x,t) + \frac{\partial P}{\partial x}(x,t) = -m(x)P(x,t),$$

with the initial condition $P(x,0) = P_0(x)$ and the boundary condition

$$P(0,t) = b(t)f \int_0^\infty h(x)P(x,t)\,dx,$$

where $b(t)$ is the parameter to be controlled. If the total fertility of women is constant and above the critical threshold

$$b^* = 1 / \left[f \int_0^\infty h(x)e^{-\int_0^x m(y)\,dy}\,dx \right],$$

then the population increases exponentially. This criterion is similar to the one obtained by Lotka with formula (10.2). Song Jian's team considered also the time-discrete version of the model, which is similar to Leslie's model (see Chapter 21). Call $P_{k,n}$ the population aged k in year n. Introduce similarly m_k, b_n and h_k. Then

$$P_{k+1,n+1} = (1 - m_k)P_{k,n}, \quad P_{0,n+1} = b_n f \sum_{k \geq 0} h_k P_{k,n}.$$

Knowing from sample surveys the mortality m_k (Fig. 25.2), the proportion of female births $f \simeq 0.487$, the age distribution of mothers h_k (Fig. 25.3), the initial condition $P_{k,0}$ which is the population's age structure in 1978 (Fig. 25.4) and varying the total fertility b (assumed constant throughout each simulation), Song Jian's team could make demographic projections for their country with a time horizon of one hundred years, from 1980 till 2080 (Fig. 25.5). Given the required thousands of additions and multiplications (year n varies between 0 and 100 years, age k between 0 and 90 years), a computer was necessary. At the time in China few people had access to such equipment except those working for the military. Song Jian, a leading expert in missile guidance, was one of them.

The projections suggested that even if China kept its 1978 fertility of $b = 2.3$ children per women, which is just above the critical threshold estimated to be $b^* = 2.19$, the population would increase from 980 million in 1980 to 2.12 billion in 2080. But China was already using almost all the land that could serve for agriculture. It even had a tendency to lose part of this land because of desertification and urbanization. How to feed such a population if progress in farm yields is not sufficient? It is the same question Malthus had considered two centuries earlier. With the 1975 fertility of $b = 3.0$, the population could even reach 4.26 billion in 2080. With $b = 2.0$, the population would reach a maximum of 1.53 billion around the year 2050 before starting to decrease slightly. With $b = 1.5$, a maximum of 1.17 billion would be reached around 2030. With $b = 1.0$, the maximum would be just 1.05 billion and would be reached around 2000. Under that assumption, the population would return to its 1978 level only by 2025.

The most surprising part of this work was its practical consequences, in fact of unequalled importance in the history of mathematical population dynamics. Indeed Li Guangyuan showed the results of the team's simulations in December 1979 dur-

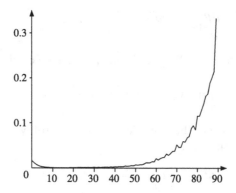

Fig. 25.2 Mortality (per year) as a function of age in 1978.

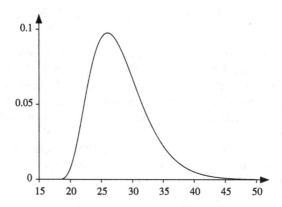

Fig. 25.3 Smoothed shape of the fertility (per year) as a function of age in 1978.

ing a symposium on population in Chengdu, Sichuan province[2]. In January 1980, Song Jian, Yu Jingyuan and Li Guangyuan published these results in a Chinese economics journal, advocating by the way a one-child policy. They also sent their article – *A report on quantitative research on the question of China's population development* – to China's top scientist Qian Xuesen, who forwarded it with recommendation to the head of the birth planning administration. The results of Song Jian's team made a deep impression on most political leaders. These were already convinced of the necessity of an increased birth control despite what Marx had written (see Chapter 5) but were still hesitating on the level of control. In February 1980, the

---

[2] Here and below, we summarize Susan Greenhalgh's detailed account [1,2].

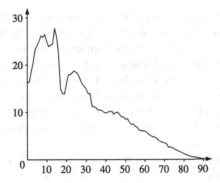

Fig. 25.4 Age pyramid in 1978. Horizontal axis: age. Vertical axis: population (in millions).

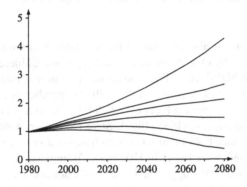

Fig. 25.5 Demographic projections (in billions) following different hypotheses on the mean number of children per woman. From bottom to top: $b = 1.0$; 1.5; 2.0; 2.3; 2.5; 3.0.

State Council and the Party's Central Committee fixed an objective for the Chinese population of 1.2 billion for the horizon 2000. In March 1980, the results of Song Jian's team were published in the *People's Daily*. In April, a commission of political leaders and population specialists examined the environmental and economic consequences of population growth and concluded that a one-child policy was necessary to reach the target set by Deng Xiaoping for the per capita income in the year 2000. The policy became official in September of that same year and an open letter explaining it to the population was published on the first page of the *People's Daily*.

By 1983, there will still many unauthorized births. It was decided that a member of each couple with already two children would be sterilized and that each forbidden pregnancy would be interrupted. However, starting in 1984, rural couples with just one daughter were allowed to have a second child. The one-child policy is still in force nowadays. Some adaptations have been introduced in recent years: if in a couple both the man and the woman were single children, then they can have two children. The repressive measures against couples having more than one child are harsh: government employees can lose their job, a costly fine has to be paid to get the administrative papers for the schooling of a second child etc. In summary, it is hard to find in the history of mathematical modelling another example with such a strong social impact. Of course the work of Song Jian and his collaborators was only one of the elements that lead to the choice of the one-child policy. But it seems to have played an important role.

As in previous chapters, the role of mathematical modelling may be a subject of concern. Starting from a real life situation, a model is built. It can be analyzed mathematically or simulated with a computer. One can then understand how the model behaves when some parameters vary. However, mathematics does not say if the model is a faithful picture of real life. Some very important aspects may have been neglected. Some models also contain an objective function, for example keeping the Chinese population under 1.2 billion. Mathematics does not say if this objective was appropriate[3].

In 1980 Song Jian was also coauthor of the new edition of the book entitled *Engineering Cybernetics* by Qian Xuesen, the "father" of the Chinese space program. He then held various high-level political positions: vice-minister and chief scientist-engineer of the Ministry of Aerospace (1981–1984), member of the Central Committee of the Chinese Communist Party (1982–2002), chairman of the State Science and Technology Commission (1985–1998), State Councilor (1986–1998) etc. He published also two other books that have been translated into English: *Population Control in China* (1985, with Tuan Chi-Hsien and Yu Jingyuan) and *Population System Control* (1988, with Yu Jingyuan). These books develop the theory of optimal control applied to population dynamics. Song Jian was elected in 1991 to the Chinese Academy of Sciences and in 1994 to the Academy of Engineers, of which he was president from 1998 till 2002.

Further reading

1. Greenhalgh, S.: Missile science, population science: The origins of China's one-child policy. *China Q.* **182**, 253–276 (2005)
2. Greenhalgh, S.: *Just One Child, Science and Policy in Deng's China*. University of California Press (2008)
3. Meadows, D.H., Meadows, D.L., Randers, J., Behrens, W.W.: *The Limits to Growth, A Report for the Club of Rome's Project on the Predicament of Mankind*, 2nd edn. Universe Books, New York (1974)

---

[3] The population in the year 2000 was estimated to be 1.264 billion. The per capita income has grown approximately from $200 to $1000 between 1980 and 2000. At the same time, the sex ratio has become extremely biased towards boys, mainly because of sex-selective abortion.

4. Song, J.: *Selected Works of J. Song*. Science Press, Beijing (1999)
5. Song, J.: Some developments in mathematical demography and their application to the People's Republic of China. *Theor. Popul. Biol.* **22**, 382–391 (1982)
6. Song, J., Yu, J.: *Population System Control*. Springer, Berlin (1988)

Chapter 26
Some contemporary problems

This chapter gives a brief overview of contemporary research on the mathematical modelling of population dynamics. The subject being quite large, only a few examples are given here of studies developed by researchers in France.

In demography a relatively new problem has appeared in the last few decades: population aging. This problem is a subject of concern not only in France (Fig. 26.1) but also in many other European countries as well as in Japan. It has important economic and social consequences: pension systems, immigration policies etc. In France, mathematical models trying to analyze the aging phenomenon are developed by the National Institute of Demographic Studies (INED) and by the National Institute of Statistics and Economic Studies (INSEE). One of the difficulties of demographic projections lies in the fact that birth rates may vary considerably over time without being foreseeable even one decade in advance. This is particularly striking if one looks back at the projections made in 1968 for the French population in 1985: these projections[1] failed to anticipate the decrease in the birth rate which occurred during the 1970s. It would be interesting to review all the predictions based on mathematical models which turned out to be wrong, especially those which found an echo in the media. This would counterbalance the impression of "progress" given by the present book, an impression which may have already appeared suspicious to the reader after reading the chapter on the Chinese one-child policy[2]. Concerning the latter subject, a new problem is now of current concern: how to soften the policy to avoid the rapid aging phenomenon expected in the next few decades. Again mathematical models contribute to the debate.

In epidemiology, among the new problems that have emerged globally in the last two decades, the development of the AIDS epidemic is particularly striking. Some models try to guess the future of the epidemic in more recently infected countries such as Russia, India or China. It is difficult to predict if the epidemic will slow down as in Western Europe and North America or if it will reach an important percentage

---

[1] See for instance the article entitled *Population (Géographie de la)* in the *Encyclopaedia Universalis*, written in 1968 and reprinted without change in later editions.

[2] This policy is often criticized in the West but seems to be relatively well accepted by many Chinese.

N. Bacaër, *A Short History of Mathematical Population Dynamics*,
DOI 10.1007/978-0-85729-115-8_26, © Springer-Verlag London Limited 2011

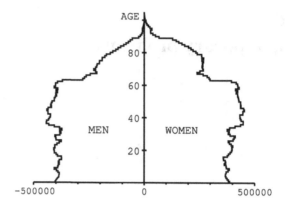

Fig. 26.1 Age pyramid of the French population on 1 January 2010. Source: www.insee.fr.

of the population as in some sub-Saharan countries. Other emerging diseases such as Ebola in Africa, West Nile fever in North America, SARS (Severe Acute Respiratory Syndrome), bird flu, chikungunya or H1N1 influenza have all been scrutinized with mathematical models, though admittedly with little success.

For SARS, one modelling difficulty was that the epidemic remained relatively limited within each country but could spread very quickly from country to country (Hong Kong and China, Singapore, Canada...). The random character of the epidemic curves in each new focus could not be neglected. As we saw in Chapters 16 and 22, stochastic models are usually more difficult to handle.

For the chikungunya epidemic that occurred between 2005 and 2006 on Reunion Island (a French overseas territory in the Indian Ocean), some models were inspired by that of Ross for malaria (see Chapter 12), the two diseases being transmitted by mosquitoes. An important aspect to take into account was the influence of seasonality. Indeed the mosquito population decreases during the southern winter, thus reducing the transmission of the disease. This can be seen in Fig. 26.2, which shows the number of new cases reported each week by a small network of about thirty general practitioners covering just a fraction of the island's population. The network did not detect any new cases during several weeks in September and October 2005, but the transmission of the disease was still continuing. Mathematical models of the epidemic were developed at the National Health and Medical Research Institute (INSERM) and at the Tropical Research Institute (IRD). Despite these models, nobody was able to foresee that the epidemic would not die out before the end of the southern winter of 2005, when it had infected just a few thousand people. Finally, almost one third of the island's population became infected, that is about 266,000 people. This shows if still necessary that predicting the future of epidemics can be quite difficult and that it is not so easy to distinguish in the early days of an epidemic if it will be a minor or a major epidemic. A parallel can be drawn with weather forecasting. This kind of forecasting relies nowadays on intensive computer simulations of complicated mathematical models of the ocean and of the atmosphere. Nevertheless predictions beyond a few days are not reliable.

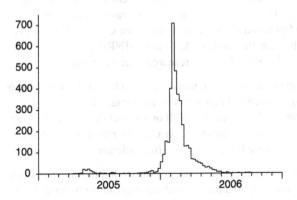

Fig. 26.2 The chikungunya epidemic in the Reunion Island in 2005–2006. Number of new cases reported per week by a small network of medical doctors as a function of time. The first small peak was reached in May 2005, the second large peak in February 2006. The numbers in this figure have to be multiplied by about 67 to get the real size of the epidemic. Source: www.invs.sante.fr.

From a more theoretical point of view, the chikungunya epidemic raised the question of how to adapt the notion of basic reproduction number $\mathscr{R}_0$ in models that assume that the environment has seasonal (e.g. periodic) fluctuations. The adaptation is not so straightforward and this raises some concern about how the parameter $\mathscr{R}_0$ has been used for other epidemics influenced by seasonality such as the 2009 H1N1 influenza pandemic.

Another problem of increasing concern that modelers have tried to analyze is that of drug resistance (antibiotics, antimalarial drugs...). Still in epidemiology, the recurrent question since the time of Daniel Bernoulli and d'Alembert of how to balance costs and benefits when the injection of a vaccine carries a potential risk is still subject to controversy and may ever remain so as the sensitivity to risk changes. Hence, following some suggestions that the vaccine against hepatitis B might cause severe complications in a very small number of cases, the French ministry of Health in 1998 stopped its vaccination campaign in schools even if the risk appeared negligible compared with that of dying after infection with the hepatitis B virus.

In ecology the study of the dynamics of fish populations still poses many problems. Nevertheless it is supposed to serve as a scientific basis for the choice of fishing quotas and other restrictions. The overfishing of the anchovy in the Bay of Biscay and of the red tuna in the Mediterranean Sea are just two recent examples. The estimation of the fish stock being often unreliable, models using such data have to be considered with caution. In France this type of study is mainly undertaken by the Research Institute for the Exploitation of the Sea (IFREMER). Some mathematical models have also played a role in past decisions of the International Whaling Commission.

In population genetics, the dispersal of genetically modified organisms is also a subject of controversy that some researchers have tried to study using "reaction–diffusion" models inspired by that of Fisher (see Chapter 20). This is the area of the National Institute for Research in Agronomy (INRA).

On the more theoretical side of research, one can mention:

- the works on partial differential equations such as diffusion equations (see Chapter 20) or age-structured equations (see Chapter 16);
- the works on stochastic models with or without the spatial dimension (see Chapters 16 and 22), including those on random networks modelling the spread of epidemics and those looking for deterministic approximations.

This type of research is mainly carried out by applied mathematicians. In recent years, several masters courses in mathematical biology have been introduced in French universities and other higher education institutions.

Like other scientific fields, the mathematical study of population dynamics is organized mainly through:

- "learned societies": Society for Mathematical Biology (since 1973), Société Francophone de Biologie Théorique (1985), Japanese Society for Mathematical Biology (1989), European Society for Mathematical and Theoretical Biology (1991) etc.
- specialized journals: *Bulletin of Mathematical Biology* (since 1939), *Mathematical Biosciences* (1967), *Journal of Mathematical Biology* (1974), *Mathematical Medicine and Biology* (1984), *Mathematical Population Studies* (1988), *Mathematical Biosciences and Engineering* (2004) etc.
- book series: *Lecture Notes in Biomathematics* (edited by Springer, 100 volumes between 1974 and 1994);
- conferences (Annual Meeting of the Society for Mathematical Biology, Mathematical and Computational Population Dynamics, European Conference on Mathematical and Theoretical Biology etc.).

Reference has been made only to the elements that claim explicitly being at the interface between mathematics and its applications to population dynamics. But for each particular area (demography, ecology, population genetics, epidemiology and so on), one can find similar elements with a variable dose of mathematical modelling.

In conclusion, the interested reader is invited to have a look at the original articles that are available on the World Wide Web. The addresses are given in the references at the end of each chapter. As Ronald Fisher once wrote about Mendel:

The History of Science has suffered greatly from the use by teachers of second-hand material and the consequent obliteration of the circumstances and the intellectual atmosphere in which the great discoveries of the past were made. A first-hand study is always instructive and often ... full of surprises.

Further reading

1. Bacaër, N.: Approximation of the basic reproduction number $\mathscr{R}_0$ for vector-borne diseases with a periodic vector population. *Bull. Math. Biol.* **69**, 1067–1091 (2007)
2. Bacaër, N., Gomes, M.G.M.: On the final size of epidemics with seasonality. *Bull. Math. Biol.* **71**, 1954–1966 (2009)
3. Bennett, J.H.: *Experiments in Plant Hybridisation*. Oliver & Boyd, Edinburgh (1965)
4. Levin, S.A.: Mathematics and biology, the interface. www.bio.vu.nl

Figures

- p. 5. Portrait by Thomas Murray (ca. 1687) held by the Royal Society in London. Chapman, S.: Edmond Halley, F.R.S. 1656–1742. *Notes Rec. R. Soc. Lond.* **12**, 168–174 (1957) ©The Royal Society.
- p. 11. Portrait by Emanuel Handmann (1753) held by the Kunstmuseum in Basel. *Leonhard Euler 1707–1783, Beiträge zu Leben und Werk.* Birkhäuser, Basel (1983)
- p. 16. Portrait once held by the Petri-Kirche, probably destroyed during the battle of Berlin in 1945. Reimer, K.F.: Johann Peter Süssmilch, seine Abstammung und Biographie. *Arch. soz. Hyg. Demogr.* **7**, 20–28 (1932)
- p. 21. Portrait by Johann Niclaus Grooth (ca. 1750–1755) held by the Natur-historisches Museum in Basel. Speiser, D.: *Die Werke von Daniel Bernoulli*, Band 2. Birkhäuser, Basel (1982)
- p. 28. Portrait by Maurice Quentin Delatour (1753) held by the Musée du Louvre in Paris.
- p. 31. Portrait by John Linnell (1833) held by Haileybury College, England. Habakkuk, H.J.: Robert Malthus, F.R.S. (1766–1834). *Notes Rec. R. Soc. Lond.* **14**, 99–108 (1959)
- p. 35. Engraving by Flameng (1850). Quetelet, A.: Pierre-François Verhulst. *Annu. Acad. R. Sci. Lett. B.-Arts Belg.* **16**, 97–124 (1850)
- p. 41. Heyde, C.C., Seneta, E.: I. J. Bienaymé, *Statistical Theory Anticipated.* Springer-Verlag, New York (1977) ©Académie des sciences, Institut de France.
- p. 45. Bateson, W.: *Mendel's Principles of Heredity.* Cambridge University Press (1913)
- p. 50. Pearson, K.: *The Life, Letters, and Labors of Francis Galton*, vol. 1. Cambridge University Press (1914)
- p. 50. Portrait of Watson in Trinity College Library, University of Cambridge. Kendall, D.G.: Branching processes since 1873. *J. Lond. Math. Soc.* **41**, 385–406 (1966)
- p. 55. Alfred J. Lotka Papers. Public Policy Papers. Department of Rare Books and Special Collections. ©Princeton University Library.

N. Bacaër, *A Short History of Mathematical Population Dynamics*,
DOI 10.1007/978-0-85729-115-8, © Springer-Verlag London Limited 2011

- p. 59. Titchmarsh, E. C.: Godfrey Harold Hardy 1877–1947. *Obit. Not. Fellows R. Soc.* **6**, 446–461 (1949)
- p. 62. Stern, C.: Wilhelm Weinberg. *Genetics* **47**, 1–5 (1962)
- p. 66. G.H.F.N.: Sir Ronald Ross 1857–1932. *Obit. Not. Fellows R. Soc.* **1**, 108–115 (1933) ©The Royal Society.
- p. 74. Whittaker, E.T.: Vito Volterra 1860–1940. *Obit. Not. Fellows R. Soc.* **3**, 690–729 (1941)
- p. 77. Yates, F., Mather, K.: Ronald Aylmer Fisher, 1890–1962. *Biog. Mem. Fellows R. Soc.* **9**, 91–120 (1963) ©The Royal Society/Godfrey Argent Studio.
- p. 81. Yates, F.: George Udny Yule. *Obit. Not. Fellows R. Soc.* **8**,308–323 (1952)
- p. 89. Heyde, C.C., Seneta, E. (eds.): *Statisticians of the Centuries.* Springer, New York (2001)
- p.97.britannica.com/EBchecked/topic/252257/J-B-S-Haldane ©Bassano and Vandyk Studios.
- p. 105. Hill, W.G.: Sewall Wright, 21 December 1889-3 March 1988. *Biog. Mem. Fellows R. Soc.* **36**, 568–579 (1990) ©Llewellyn Studios, Chicago.
- p. 101. Nybølle, H.C.: Agner Krarup Erlang f. 1. Januar 1878 - d. 3. Februar 1929. *Mat. Tidsskr. B*, 32–36 (1929)
- p. 113. Tikhomirov, V.M.: A.N. Kolmogorov. In: Zdravkovska, S., Duren, P.L. (eds.) *Golden Years of Moscow Mathematics*, 2nd edn., pp. 101–128. American Mathematical Society (2007)
- p. 113. *I. G. Petrowsky Selected Works Part I.* Gordon and Breach, Amsterdam (1996) ©Taylor and Francis Books UK.
- p. 117. Photograph by Denys Kempson. Crowcroft, P.: *Elton's Ecologists, a History of the Bureau of Animal Population.* University of Chicago Press (1991)
- p. 121. ©Geoffrey Grimmett.
- p. 127. Charlesworth, B., Harvey, P.: John Maynard Smith, 6 January 1920–19 April 2004. *Biog. Mem. Fellows R. Soc.* **51**, 253–265 (2005) ©The Royal Society.
- p. 133. ©Samuel Schlaefli / ETH Zürich.
- p. 141. Selected works of J. Song. *Science Press*, Beijing (1999) ©Song Jian.

Index

N. Bacaër, *A Short History of Mathematical Population Dynamics*,
DOI 10.1007/978-0-85729-115-8, © Springer-Verlag London Limited 2011